WILD NEW WORLD

This book is published to accompany the television series *Wild New World*, first broadcast on BBC2 in 2002 and in the United States on the Discovery Channel in 2003.
Executive Producer: Neil Nightingale. Series Producer: Miles Barton.

Published by BBC Worldwide Ltd, Woodlands, 80 Wood Lane, London W12 0TT.
Published in the United States by Yale University Press.

ISBN 0-300-09819-7

Library of Congress Control Number: 2002109289

A catalogue record for this book is available from the British Library.

Editorial Director: Shirley Patton
Commissioning Editor/Copy Editor: Joanne Osborn
Project Editor: Martin Redfern
Book Art Direction and Design: Lisa Pettibone
Cartographer: Oliver Pearson
Production Controller: Kenneth McKay

Set in Sackers, Adobe Garamond and Helvetica
Printed and bound in Great Britain by Butler & Tanner Ltd, Frome and London
Colour separations by Radstock Reproductions Ltd, Midsomer Norton
Jacket printed by Lawrence Allen Ltd, Weston-super-Mare

The paper in this book meets the guidelines for permanence and durability of the Committee on Production Guidelines for Book Longevity of the Council on Library Resources.

10 9 8 7 6 5 4 3 2 1

PICTURE CREDITS
The publishers would like to thank the following for providing photographs and for permission to reproduce copyright material. While every effort has been made to trace and acknowledge all copyright holders, we would like to apologize should there have been any errors or omissions.

AKG (London) Ltd 179; B. & C. Alexander 11br; Deborah Allen 161br, 188l; Ardea 176-7 (Donald Burgess); Miles Barton 3, 10, 16, 24, 34, 73tr; Nigel Bean 1, 4, 11tc, 25, 28, 38-9, 73br, 79, 89b, 100, 101tc, 102, 109br, 112b, 118, 122,124-5; John Brown 86, 112t; Corbis 13 (Steve Kaufman), 43tc (Darrell Gulin), 119l (Phil Schermeister), 127tr (Richard Cummins), 128 (Dean Conger), 132 (Kevin Fleming), 173l, 178, 186-7 (Paul A. Souders),188r; Courtesy of the Burton Historical Collection, Detroit Public Library 181r; Stephen Dunleavy 30, 44, 46, 47t, 48-9, 64, 65,164; FLPA 26l (Terry Whitaker), 39r (Helen Rhode), 54 (Mark Newman), 59t (Terry Whitaker), 62 (Jurgen & Christine Sohns), 73bc (Mark Newman), 78 (Leonard Lee Rue), 81 (S. Maslowski), 110-1 (Mark Newman), 116 (Martin Withers), 119r (David Hosking), 187r (S. Maslowski); Getty Images 66 (Chip Porter); Ian Gray 72, 97; Dale Guthrie 22, 31; Robert Harding Picture Library 160 (Simon Harris),185 (Louise Murray); Courtesy of The Mammoth Site, Hot Springs, South Dakota 84l; John Hyde, Wild Things 43bc, 60; Courtesy Illinois State Museum 51 (Martin Roos); University of Michigan, Museum of Palaeontology, Ann Arbor Michigan 173r; Minden Pictures 76-7 (Jim Brandenburg), 184l (Jim Brandenburg); NASA 137; Natural History Museum, London 52; Nature Picture Library 11tr (Paul N. Johnson), 14 (Staffan Widstrand), 26r (Paul N. Johnson), 32 (T. Andrewartha), 61 (Mike Salisbury), 63 (Neil Lucas), 68t (Jeff Foott), 84r (Richard du Toit), 92r (Peter Blackwell), 107 (Pete Oxford), 125 (Mary Ann McDonald), 126-7 (Nigel Bean), 127b (Nigel Bean), 134-5 (Jeff Foott), 141 (Jeff Foott), 144 (Nick Gordon), 146 (Lynn M. Stone), 148 (Staffan Widstrand), 149l (Torsten Brehm), 149r (Nick Gordon), 161tr (Lynn M. Stone), 177r (Adam White) 180-1 (Lynn M. Stone); NHPA 15 (Stephen Krasemann), 39c (John Shaw), 42 (John Shaw), 75 (Rod Planck), 90 (Martin Harvey), 96 (Stephen Dalton), 101r (Andy Rouse), 104l (T. Kitchin & V. Hurst), 109tr (Jany Sauvanet), 133r (T. Kitchin & V. Hurst), 136 (James Carmichael Jr), 138-9 (James Carmichael Jr); OSF 43r (David Boyle/ Animals Animals), 47b (Norbert Rosing), 68b (Jeff Foott/ Okapia), 89t (Judd Cooney), 101bc (Stan Osolinski), 105 (James H. Robinson), 120l (Paul Berquist/Animals Animals), 124l (Sean Morris), 153 (Zig Leszczynski), 183 (Joe McDonald/Animals Animals); Courtesy of the George C. Page Museum, Los Angeles 59b, 93, 95r (Ed Ikuta); Science Photo Library 174; The University of Texas at Austin, from the Collections of the Texas Memorial Museum of Science and History 56c (© Glen Evans 2002); Vireo 115 (S. Lafrance); Adam White 7, 29r, 133tc, 133 bc, 143t, 147, 150, 152, 157, 167.

Digital images created by BBC MediaArc © BBC Worldwide 2002

PREHISTORIC AMERICA

A JOURNEY THROUGH THE ICE AGE AND BEYOND

Miles Barton, Nigel Bean, Stephen Dunleavy,
Ian Gray, Adam White

Foreword by D. Bruce Means

Yale University Press
New Haven and London

CONTENTS

FOREWORD

Growing up in Alaska, I developed a keen appreciation for big animals. Bull moose were a familiar sight around the cabin where I grew up, and how can I forget being charged by a huge grizzly bear one midsummer evening? In my college years, I spent the summers picking through the discarded remains of Pleistocene animals, left behind by gold miners who had hydraulically sluiced bones out of frozen gravels north of Fairbanks.

Today, a whole continent away in Florida, the skulls of an extinct Alaskan horse and giant bison sit on a shelf in my office next to the tusk of a small woolly mammoth. All around the room lie the bones of other ancient creatures, stained black by time and tannins. Skulls of giant ground sloths, molars of mastodons and Columbian mammoths, and the bones of dozens of creatures that no longer exist in North America: giant beavers, llamas, tapir, glyptodonts, dire wolves, sabre-toothed cats and a giant tortoise larger than any alive today.

I discovered these bones underwater, in the bottom of Florida's spring-fed rivers, where the ancient megafauna came to drink and died, or were slain. Some of the bones have clear butcher marks – obviously these animals had come headlong into contact with humans.

Many nights alone in my office, I have tried to conjure up images of what these splendid animals looked like and how they might have lived and died. I have often thought, 'I wish I could go back in time and see them in the flesh.' And now I can. The authors of this book – and the six-part television series that it accompanies – have brought the animals to life with amazing computer graphics. Now we can actually *see* woolly mammoths and giant short-faced bears standing in the stunning scenery of my youth. The thrill of finding and studying bones has been fanned into new flames of excitement for me. No other production, written or televised, has so masterfully or beautifully presented this story. As a vertebrate biologist, I tip my hat to the authors and the BBC for a job superbly done. I also salute the Discovery Channel and Yale University Press for bringing this fine project to the United States.

D. BRUCE MEANS
Department of Biological Science, Florida State University

Many of Florida's springs are filled with ancient fossils, which provide a fascinating glimpse of Ice-Age North America.

INTRODUCTION

Flying over the ice fields of Canada it is easy to imagine being back in the last Ice Age. There is ice as far as the eye can see. Glaciers roll down the valleys, towering ice sculptures rise out of the mountainsides, and exquisite turquoise pools glisten in the fissures below. Faced with a herd of shaggy muskoxen, their massive horns confronting danger and their long coats blowing in the icy wind, it doesn't take much to believe there might be a woolly mammoth over the next ridge. Just think if we could go back in time to an age when humans first encountered this dramatic landscape.

The vast continent of North America is truly the New World. The exact timing of the first people's arrival and how they got there is a highly contentious issue and the arguments would fill another book. What is known is that there were people throughout the continent 13,000 years ago – the Clovis people, so-named after a town in New Mexico where the first of their uniquely shaped fluted spear points were found. Many now believe that there were people in North America well before 13,000 years ago, and it is likely that there was more than one wave of colonization.

But whenever and however the first people arrived, our interest lies in what they would have seen in this new land. They would have encountered creatures they had never come across before – ground sloths and glyptodonts, giant short-faced bears and mastodons. When you add to this list camels and horses, sabre- and scimitar-toothed cats, dire wolves, lions and cheetahs, it is no wonder that Ice-Age North America has been described as an American Serengeti, teeming with a variety of large mammals that rivals the East African plains of today.

The idea of going back in time to follow in the footsteps of the first Americans is a tantalizing one. There is a wealth of fossil evidence on the continent's pre-historic wildlife. With modern computer techniques it is now possible to flesh out these ancient bones, put fur on them and indeed bring them to life on screen. This was the premise for a new kind of natural-history television series – to build a picture of what those first people would have seen by combining real animals of the present with computer-generated creatures of the past.

This is a fascinating detective story with a remarkable cast of characters, including Blue Babe, the extinct steppe bison, covered in frozen silt for over 30,000 years, and the Manis mastodon, which died in mysterious circumstances in Washington 14,000 years ago. Natural Trap Cave is another amazing link to the past: it has been a pitfall trap for animals for 20,000 years and its walls bear the scars of their futile scrabblings to escape. Finally there is the young glyptodont – a strange cross between an armadillo and a tortoise – whose skull shows signs that it was killed by a deadly feline predator. By investigating these and other stories, my fellow authors and I hope to offer a unique glimpse into the past. Our colleagues in BBC MediaArc have helped to complete the picture by bringing these extinct creatures to life in a highly realistic manner.

At the beginning of a new millennium it is worth reflecting on a time, not so long ago, when an entire continent was free of humankind. It may be just coincidence but as many as 80 per cent of the large animals disappeared within about 1000 years of the first people's arrival in North America. It seems likely that our ancestors had a hand in the extinction of many of these extraordinary beasts. It is our loss that we will never again see woolly mammoths striding through the desolate steppes, nor stare in awe at the mighty Columbian mammoth, far bigger than any elephant alive today. And we will never again be thrilled by the power of a sabre-toothed cat, nor marvel at the sight of the lumbering glyptodont. Let us hope that today's megafauna – the elephant and the rhinoceros, the tiger and the panda – do not have to be brought back to life electronically for our descendants to meet them in the future.

MILES BARTON
Series Producer

BERINGIA

Left: At 5959 m (19,550 feet), Mount Logan is the highest mountain in Canada. **Above centre:** Przewalski's horses most closely resemble the prehistoric wild horses of Beringia. **Above right:** The bizarre-looking saiga antelope once roamed Ice-Age Beringia. **Below right:** The tundra of the northern Yukon resembles the landscape that the first people would have seen as they arrived in North America.

ALASKA AND THE YUKON TODAY

The far northwest of the American continent is a vast and remote land with many relics of the last Ice Age still to be discovered in its landscape. The region covering Alaska and the Yukon territory in Canada is made up of several great wildernesses. To the south, the forests are dominated by evergreen Sitka spruce, with small groves of birch and poplar providing a riot of colour in autumn. In the centre, the great Alaska Range includes Mount McKinley – at 6200 m (20,341 feet) the highest mountain in North America – with a mass of glaciers in its shadow. The ice sheets here are sculpted by the elements into strange rivulets and towers of brilliant white and ultramarine, while below the icy plateaux glacier-fed lakes gleam like turquoise gems in the lee of the mountains. Further north are the 'barren grounds' where trees do not grow and the earth a metre or so below the surface is always frozen. This permafrost keeps water at surface-level and so, despite low rainfall, the ground is boggy and dotted with pools.

The climate of the region is as extreme as the landscape. During the brief summer, from June to August, the sun shines for up to 24 hours a day and temperatures can exceed 30°C (86°F). However, when winter returns in late August the days again become long and dark; temperatures can drop to -50°C (-58°F) with a wind chill making it even colder. Fairbanks in the Alaskan interior has one of the widest temperature ranges of any city in the world. The temperature can drop to -55°C (-67°F) in January and rise to 37°C (99°F) in July.

Despite these apparently inhospitable conditions, an array of big game has survived here, from moose to grizzly bears and muskoxen to wolves. Not so very long ago, during the last Ice Age, the region was also home to monstrous prehistoric creatures and was probably the point of arrival for humans on the continent.

So what would the first people to set foot here have seen? The evidence lies beneath the frozen ground, and in the wildlife and the landscape of today. By following these clues we can build a picture of how the far northwest would have looked and how its inhabitants would have lived around 14,000 years ago.

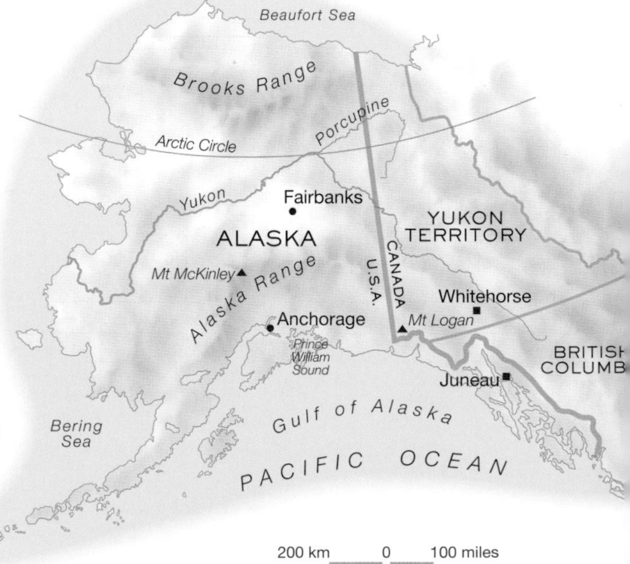

Mount McKinley dominates the beautiful wilderness of Denali National Park, Alaska.

MEGAFAUNA IN THE PRESENT DAY

Many of the species the first people encountered are still present today. There are about 1 million caribou in Alaska and the Yukon, and the western Arctic herd alone numbers a staggering 500,000. Caribou of these northern herds, including the Porcupine herd on the Alaska–Yukon border, form the most impressive gathering of large mammals anywhere in North America, rivalling the wildebeest migrations of East Africa. Caribou are perfectly designed to roam this vast landscape. Their large concave hooves spread widely to support the animals on both soft snow and tundra, and the hooves are broad enough to function as paddles for swimming rivers. The big herds are always on the move in search of their diet of lichens, willows and sedges. They can cover up to 80 km (50 miles) a day and will travel a total distance of 650 km (400 miles) between their summer and winter feeding grounds.

The bull muskox has an impressive buttress at the centre of its horns. This protects its brain as it head butts its rivals.

Caribou cope with the harsh living conditions by constantly moving from place to place. Muskoxen do just the opposite – they have been described as the couch potatoes of the Arctic. In winter they lower their metabolic rate and stay on windy hilltops or ridges where the snow is thinner and they can reach the sedges, moss and willow twigs that will sustain them until the summer. Their remarkable coat consists of an outer layer of shaggy hair up to 60 cm (2 feet) long, which almost reaches the ground, making it look as if the animal is wearing a skirt. In fact it is the short, soft underwool (or qiviut) that really keeps muskoxen warm in winter. It is eight times as warm as sheep's wool and the most effective insulator of any animal fur.

With their shaggy coats and massive heads and horns, muskoxen seem huge. In fact, a full-grown bull measures only 1.4 m (4.6 feet) at the shoulder, and the average weight of a male muskox is 340 kg (750 pounds) and of a female 250 kg (550 pounds).

Muskoxen live in herds of between 10 and 40 animals, made up mainly of cows and their offspring. The breeding season in late summer is the one time when male muskoxen become animated as they compete for access to females. The senior bull drives young pretenders away, engaging in dramatic head-to-head clashes with its rivals. It has been calculated that the force of two muskoxen clashing is equivalent to that of a car ramming a concrete wall at 27 km/h (17 mph). On a calm day the sound can be heard up to 1 mile (1.6 km) away.

Typically, muskoxen's response to danger involves the minimum of effort. They simply bunch together with their massive horns facing the threat. From the rear, muskoxen are vulnerable, but from the front a predator faces a row of deadly horns. This defence was effective against wolves and earlier Ice-Age predators but was useless against people, armed with spears and guns, who could pick off the whole herd. Muskoxen numbers declined until well into the twentieth century, with the animal becoming extinct in Alaska, but following their protection and reintroduction to parts of their former range, the numbers are now on the increase.

The muskox has a prehistoric, Ice-Age appearance and indeed it has lived in the region for tens of thousands of years. However, other creatures that once roamed the region have since disappeared. There was also a land that has now vanished – Beringia. This once joined Alaska to Siberia and owed its very existence to the last Ice Age.

CARIBOU ICE PATCHES

Caribou are plagued by mosquitoes, so in the summer they gather together in their hundreds in areas where the winter ice and snow do not melt and mosquitoes are less active. These same patches have been used by caribou for thousands of years.

Recently, pits have been dug into some of these ancient patches in the Yukon. Deep below the surface, scientists have found frozen dung up to 9000 years old and they expect to discover even older sites. By analyzing the dung they can discover what the caribou ate and therefore what vegetation was present at that time.

Even more remarkable is the discovery of human artefacts such as spear points and parts of spear throwers or atlatls (see Chapter 6, p. 166). These provide clues as to how the first people in the region may have hunted. Then, as now, the caribou and their calves, huddling in these small patches for relief from both the insects and the summer heat, would have been obvious targets for hunters.

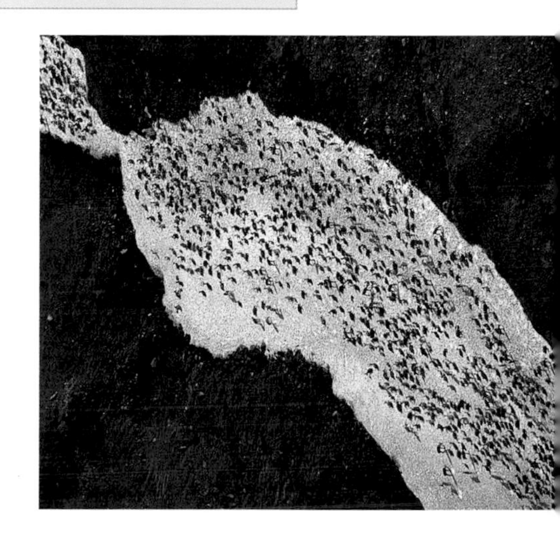

THE LAST ICE AGE

Ice Ages have occurred regularly during the last 2 million years of the world's history. They are the product of a 100,000-year cycle in climate, known as the Milankovitch cycle, which is created by the behaviour of the Earth in its orbit around the Sun. Over time, variations occur in the length of the orbit's ellipse, taking the Earth nearer to or further from the Sun. There are also variations in the tilt of the Earth's axis in relation to the Sun. The key result of these variations is the contrast between winter and summer temperatures. At one extreme of the cycle the summers are hot enough to melt the ice from the preceding winter; at the other extreme they are not, and so ice accumulates year by year to form sheets. The result is an Ice Age.

The last great Ice Age, called the Wisconsinan in North America, occurred during the late Pleistocene era between 90,000 and 10,000 years ago. It peaked about 20,000 years ago. The Wisconsinan glaciation gets its name from the state (Wisconsin) where the ice made its furthest extent south and where evidence of this glaciation (in the form of rocks dragged and left behind by the ice) was first found. At the peak of this Ice Age up to half the continent was under glaciers. The great ice sheets of North America were the vast Laurentide sheet, which spread from the northeastern Arctic down to the middle of the continent, and the smaller Cordilleran sheet, which spread from the western mountains.

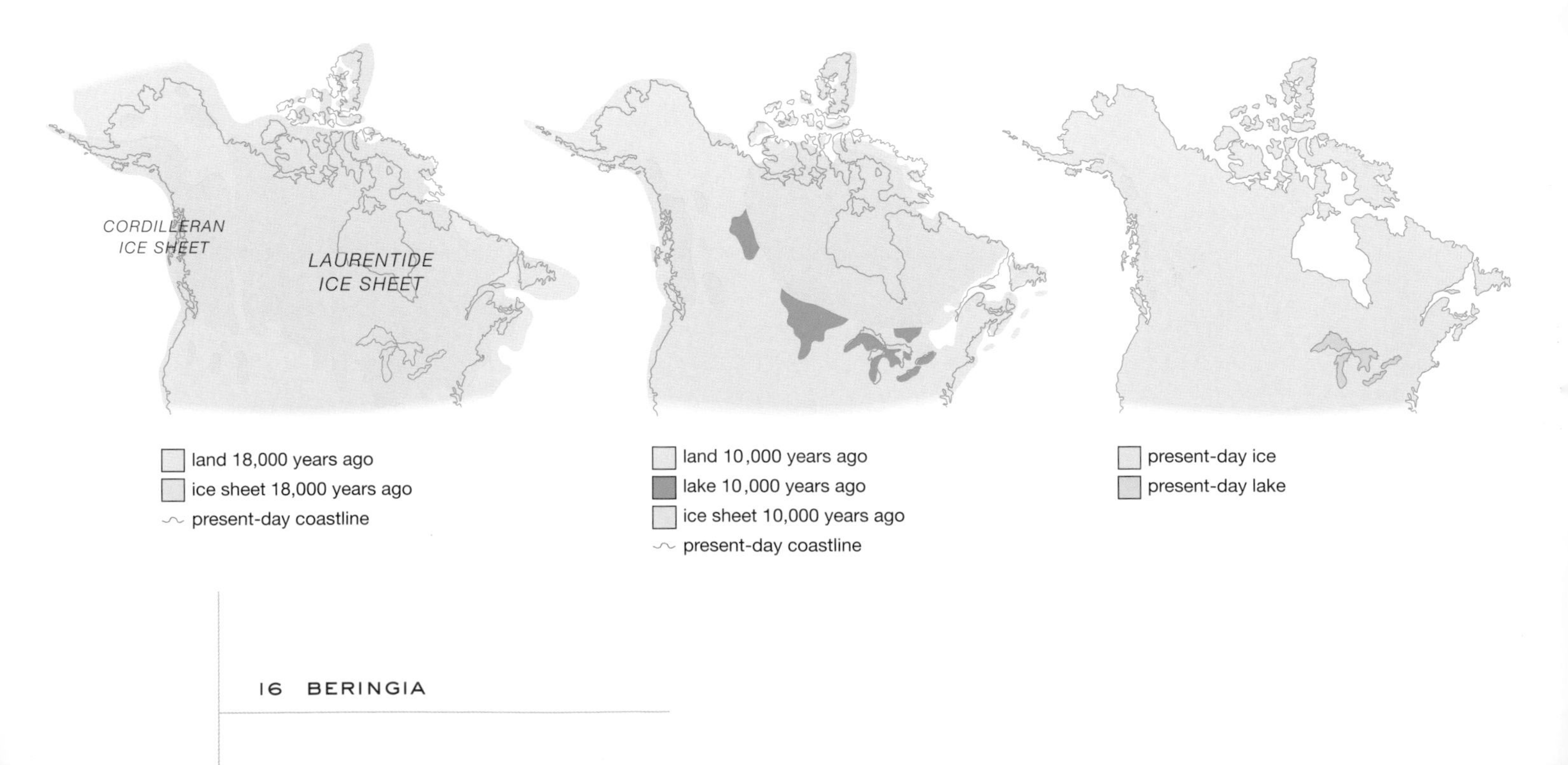

The Wrangell–Saint Elias ice field gives us a good idea of how nearly half the continent would have looked during the Ice Age. Only the tops of the mountains project above the ice sheets, which are up to 1.6 km (1 mile) thick.

ISLANDS IN THE ICE

The ice sheets of the Wrangell–Saint Elias ice field are broken in places by mountain peaks protruding from beneath the surface. These spikes of rock are known as 'nunataks' and many have remained unchanged since the Ice Age. Tiny meadows, only a few metres square, survive on their sunnier slopes, and grasses and plants, such as Jacob's ladder and saxifrage, cling to the rocky outcrops. More remarkable still, these peaks are visited by bees, crane flies and other insects. Rosy finches, water pipits and ravens can also be found. However, the most surprising inhabitants are mammals such as the pika, a tiny relative of the rabbit, which manages to colonize even the remotest of peaks.

Even today there are places where you can still see an authentic Ice-Age landscape. The Wrangell–Saint Elias ice field on the Alaska–Yukon border is a good example. The last remaining part of the great Cordilleran ice sheet, it is the largest non-polar ice field in the world and shows what much of the continent would have looked like at the height of the glaciation around 20,000 years ago.

Sheets of ice stretch as far as the eye can see, with strange shell-like patterns scalloped into the surface. Snow clings to mountainsides in great crumbling chunks while in the glaciers below, ultramarine pools glint in the sunlight. Rivers run across this glacial landscape and suddenly disappear through the ice to the valleys below. The ice here is up to 900 m (2950 feet) deep and the glaciers move up to 200 m (660 feet) a year as they grind and sculpt the landscape around them. Evidence of their work can be seen to this day far to the south in places such as Banff and Waterton–Glacier National Parks in the heart of the northern Rockies. During the last Ice Age, glaciers also radically changed the north of the continent, leading to the human invasion of North America through the creation of the Bering land bridge.

THE FORMATION OF THE BERING LAND BRIDGE

The vast glaciers that formed in the northern hemisphere locked up much of the world's water as ice. Global sea levels dropped by as much as 100 m (330 feet) as a result. As in earlier Ice Ages, this exposed the floor of the Bering Sea and created a land connection between Alaska and Siberia. At the peak of the last Ice Age the land bridge was 1600 km (1000 miles) wide and would have covered an area twice the size of Texas. For the first time in about 100,000 years, animals could once again travel across the land bridge from Siberia into the North American continent.

The land bridge was part of a larger ice-free area called Beringia, which included Siberia, Alaska and parts of the Yukon. Beringia was bounded by the then permanently frozen Arctic Ocean and the continental ice sheets. Rain and snow tended to fall on the high southern ice fields of the Yukon and Alaska, thus reducing the amount that fell on the Beringian side. At the height of glaciation the retreat of the sea meant that most of the land was far from maritime influence and so had an arid, continental climate. The low winter snowfall prevented glaciers from forming and left grass and other vegetation accessible to grazers throughout the winter. This is what made Beringia habitable at a time when much of the land to the south was buried in ice.

THE MAMMOTH STEPPE

As well as creating a dry climate, the ice sheets also made loess – a fine dust produced by the grinding action of the glaciers and deposited on the edge of streams emerging from the ice front. Loess blew across Beringia, establishing a well-draining soil. The result was a land of grassy steppes. An array of tiny plants including grasses, sedges, herbs, dwarf birch and willow provided a highly nutritious rangeland capable of supporting the giants of the past. Grass grows from its base and can easily cope with its stem tips being constantly nipped off. It is also more nutritious and less toxic than other plants. Its only defence is the amount of silica in its blades, which wears down herbivores' teeth over time.

This mixture of steppe and tundra plants was unlike the tundra or boggy muskeg found in the region today. Scientists have coined the term 'Mammoth Steppe' (after the enormous herbivore) to describe this unique environment. Some even believe that the grazing action of these massive beasts maintained the grassy landscape, which subsequently disappeared due to the extinction of the megafauna, rather than the other way round. Whatever the reason, Beringia's most impressive inhabitant was of course the woolly mammoth.

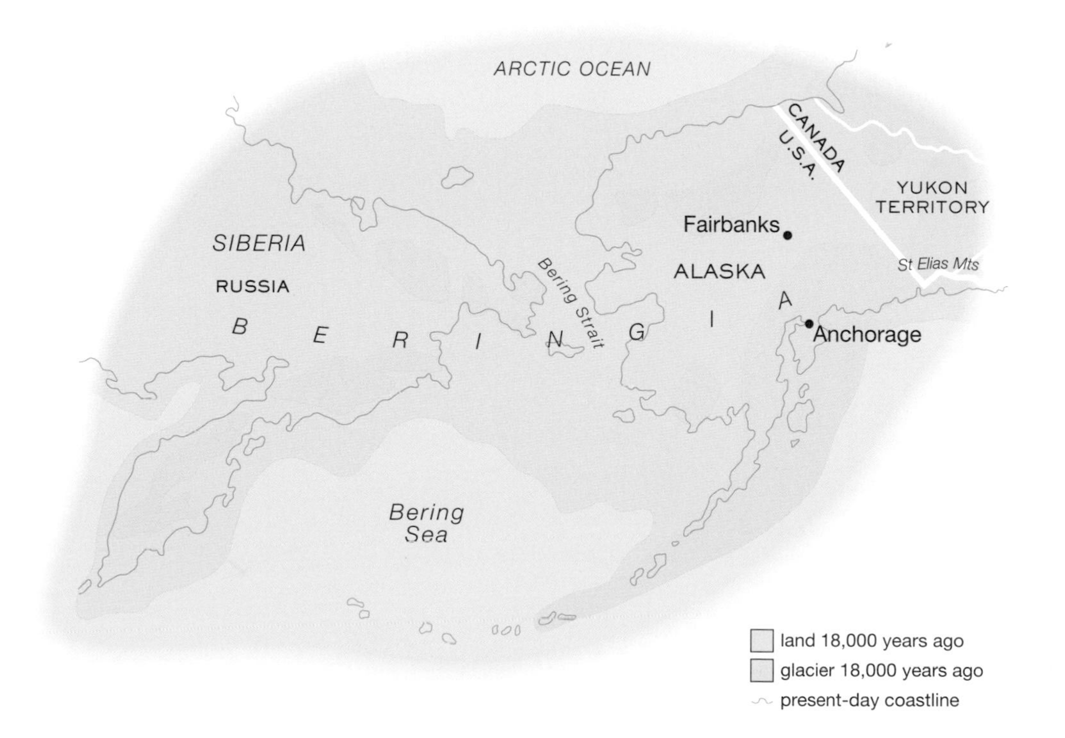

WOOLLY MAMMOTH

KEY FACTS

Common name: Woolly mammoth

Scientific name: *Mammuthus primigenius*

Size: 3 m (10 feet) tall at the shoulder

Weight: 4535 kg (4.5 tons)

Diet: Vegetarian – grass, dwarf willows and sedges

Habitat: Steppes

The woolly mammoth evolved as a species in Siberia between 300,000 and 200,000 years ago and travelled across the Bering land bridge soon afterwards. An adult male would have stood 2.8–3.4 m (9–11 feet) at the shoulder and weighed up to 6000 kg (6 tons) – about the same size as an Asian elephant. Females were smaller and lighter. The woolly mammoth's most obvious feature was its shaggy coat. It differed from modern elephants by also having huge helically curved tusks, a knob-like dome on its head, and a sloping back with a hump.

This impressive beast was well adapted to the cold. Its coarse outer coat was made up of dark brown hair, up to 90 cm (3 feet) long on certain parts of the body. Each hair, which was tough and springy, was about six times thicker than a human hair. The long hair on the animal's flanks and belly created a skirt that resembled the muskox's, and the woolly mammoth probably moulted its coat each summer in the same way. Relatively small ears and a short trunk helped to conserve heat, and beneath the skin was a 10 cm (4 inch) thick layer of subcutaneous fat for further insulation.

The tusks of male woolly mammoths were larger than female tusks, indicating that they were used by bulls to establish their dominance over rivals. Males would sometimes have competed violently for females, tussling and pushing with these enormous weapons. The largest tusk ever found was 4.2 m (13.7 feet) long and weighed 84 kg (185 pounds). Typical male tusks were about 2.5 m (8 feet) long and weighed 45 kg (100 pounds). Female tusks were less than 2 m (6.5 feet) long and weighed only 10 kg (22 pounds). The tusks grew in a complex helical shape and may have been used to clear snow in order to reach the grass beneath or to break ice on streams to reach water during the winter.

The enormous teeth of woolly mammoths give us clues about their diet. They had a series of up to 26 enamel ridges, serrated to form a washboard-like surface. This was particularly effective in grinding down grasses to extract the maximum nutrition. Woolly mammoths were adapted for grazing: this is confirmed by an investigation of the stomachs of frozen mammoths found in Siberia. Up to 90 per cent of the stomachs' contents were grasses, along with a few twigs of willow and larch. The mammoth's delicate trunk, with its two finger-like projections at the tip, was ideal for plucking the short grass of the Mammoth Steppe. These huge animals needed to eat up to 180 kg (400 pounds) a day and so they had to spend up to 20 hours of each day feeding. They were always on the move in search of fresh grazing land. The mammoths' constant cropping of the short grass helped to maintain the unique landscape of the Mammoth Steppe, which no longer exists today.

Female woolly mammoths probably lived in family groups of between two and nine animals, led by the eldest matriarch.

Woolly mammoths were social animals with a matriarchal family life similar to that of elephants. Young mammoths were cared for by older sisters and aunts and were the centre of attention in the herd.

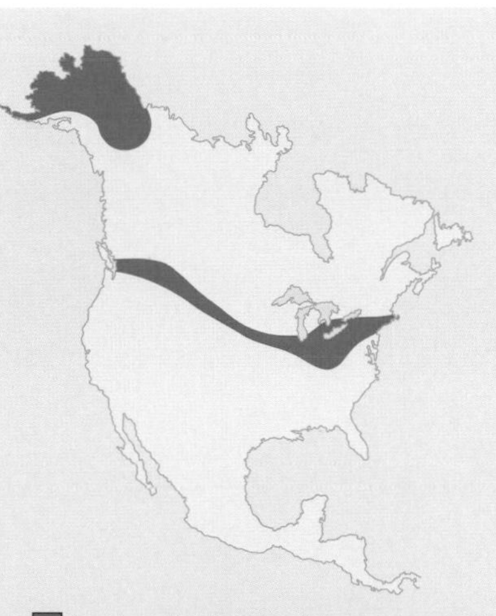

Males led a more solitary existence, probably only joining the females to breed. African elephants often time their birth season to coincide with the rains, when there is plenty of new grass for the mothers to convert into milk for their young. In the same way, young woolly mammoths may have been born in the spring to coincide with the growth of spring grass. Assuming a gestation period of 22 months (that of an elephant), the breeding season would have been during the late summer.

Like elephants, adult woolly mammoths are unlikely to have had any predators except human hunters. However, young mammoths were sometimes killed by scimitar-toothed cats and American lions. Recent research on tooth development in mammoths indicates that they may have lived longer than elephants – perhaps up to 80 years old.

Even today, placer miners in Alaska and the Yukon still mine for gold by thawing the permafrost with high-pressure hoses. Many important prehistoric discoveries have been unearthed in this way.

MAMMOTHS, MUCK AND GOLD

Much of our knowledge about the ancient beasts of the region has been discovered through mining. The gold rushes of Alaska and the Yukon at the end of the nineteenth century have played a major part in revealing the secrets of the past. Although the miners did not always get rich, they did discover a treasure trove of prehistoric evidence.

Some of this evidence was in the form of conventional fossilized bones. Mammoth tusks too have survived intact and indeed have been traded as ivory right up to the present day. Thousands of years after the mammoths' extinction these ancient beasts are still being exploited for their ivory.

There was even more valuable evidence in the frozen silt, or muck, of the north. The frozen mummified remains of some prehistoric beasts have been so well preserved that we can even tell what their last meal consisted of. The reason for this array of ancient specimens in the Alaska and Yukon region is its particular climate and geology at the end of the last Ice Age. Each summer, soils on the hills would thaw a few centimetres in depth and slide down the slopes in thick, slow-moving waves, rather like melted chocolate over frozen ice cream. As they oozed along, these waves would slowly pick up the remains of plants and animals, which built up in the valley bottoms. The result was the accumulation of dark organic frozen silt, containing many fossils well preserved by the cold.

This muck lay there for thousands of years until placer miners, looking for gold deposits in the ancient stream beds, blasted it away with their powerful hose jets. And the rest is history.

EFFIE, THE BABY WOOLLY MAMMOTH

Effie is the best-preserved woolly mammoth found in North America. Discovered in 1948, he or she was washed out of the muck of a gold-mine in a creek near Fairbanks in Alaska. From carbon dating, this very young mammoth is believed to be 21,300 years old. Only the head, foreleg and shoulder were found, but the skin and muscle are so well preserved that they have been used for DNA studies of mammoth evolution. The mammoth was nicknamed Effie after the Fairbanks Exploration (FE) branch of the US Smelting, Refining and Mining Company.

Effie's small size indicates that he or she was under a year old at the time of death. Young mammoths, like elephants, are believed to have been closely guarded by the mother and the herd until they were at least two years old and so it is unlikely that a predator was involved in the death. Perhaps Effie's mother failed to produce enough milk for her infant during the long winter and he or she starved. Or perhaps Effie fell off a cliff or into some mud and got stuck. Whatever the reason, Effie's mummified remains were found in several parts (with the tip of the trunk bitten off) and it seems that the body had been scavenged. What was left was fortuitously buried in silt, probably during the summer melt, and remained there for over 21,000 years.

MAMMOTHS IN LEGEND

Long before miners arrived, the indigenous people of the region knew all about the ancient remains of these monsters. They featured in a number of local legends. The KoYukon people of the lower Yukon River believed that the souls of the dead journeyed upriver to near what became the gold town of Dawson. There they waited to be reborn. In the afterworld they continued hunting and fishing, much as they had in life. The animals they hunted were called the 'Underground Game' and their bones could sometimes be seen protruding from the muck.

The Vuntut Gwitch'in people tell of mammoths that used to live in the ground around the Old Crow River. They believe that, when nearing death, the animals pushed up through the ground to reach the river. In their view, this explains why the bones of these animals are so often found protruding from river banks.

BERINGIA'S BIG GAME

A whole menagerie of beasts roamed the steppes alongside the woolly mammoth. Steppe bison were probably the most common, followed by the wild horse, the muskox and caribou.

The wild horse was widespread throughout North America during the last Ice Age. In fact, the very first horses evolved on the North American continent around 55 million years ago. Over millions of years they slowly extended their range to the grasslands of most continents on Earth. The wild horses of Beringia were probably similar in appearance and behaviour to modern-day Przewalski's horses

Przewalski's horse is a native of the wind-blown grassy steppes of Mongolia, where it evolved to cope with similar conditions to those found in Beringia. Sadly, until recently, it had last been seen in the wild in 1968, although it has been successfully bred in captivity for a century. In the 1990s it was finally released back into the wild in the Hustain Nuruu reserve in Mongolia. Recent studies of the horse's behaviour under free-range and wild conditions give us clues to how the original wild horses in Beringia may have lived. Przewalski's horses live in small groups of about 10 animals, consisting of a stallion, several mares and their young. There are strong bonds between the individuals in the harem and they act together when threatened by wolves. The mares will typically gather around their young, with their hind legs ready to kick out at the predators. The stallion then circles the group and attacks the wolves directly.

Left: Przewalski's horses grow thick coats and long manes in the winter to cope with the cold on the steppes.
Right: Today, Przewalski's horses are bred in captivity all over the world, but only a handful have been returned to the wild.

Left: The impressive European bison or wisent resembles the extinct steppe bison more closely than the more familiar North American plains bison. **Right:** The saiga antelope's extraordinary nose is an adaptation to the cold, dry conditions on the steppes.

Fossil evidence suggests that both Dall's sheep and muskoxen were present during the last Ice Age, as they are today. Then, Dall's sheep were not just confined to the mountains but grazed the wind-blown plateaux too. Like woolly mammoths, muskoxen originally evolved in Asia and travelled to North America via the Bering land bridge. In addition to the tundra muskox there was the larger helmeted muskox, which was nearly as big as a bison.

Another Asian creature that arrived via the land bridge was the saiga antelope. In 1976, fossil specimens of these strange looking creatures were recovered from a 13,000-year-old deposit at the Bluefish Caves in northern Yukon. Saiga antelopes are thought to have disappeared from North America at the end of the Ice Age and they are now confined to central Asia, where huge herds can still be found.

The most bizarre feature of the saiga is its enormous nose, which makes it look as if the head of an elephant seal has been grafted onto a small deer. The large nasal cavity pre-warms cold air being breathed in and reduces the loss of water on exhalation. This allows the saiga to breathe efficiently in cold, arid conditions – similar to those that once prevailed in the Mammoth Steppe. The saiga are small and light, weighing an average of 30 kg (66 pounds). They are also swift animals, capable of speeds in excess of 70 km/h (43 mph). Their pacing gait indicates that they are only able to live on flat terrain.

The steppe bison was a slightly larger animal than the modern bison. It had longer horns with tips that curved forward. The horns of large males had a spread of about 1 m (3.3 feet). Steppe bison may have resembled the forest-dwelling European bison or wisent more closely, but evidence from the skull and teeth structure and the contents of mummified carcasses show that they were grazers like the plains bison of today.

The presence of these grazers in the region confirms that the vegetation was once dominated by grasses. The landscape of Beringia resembled the Asian grassland steppes and it supported far more herbivores than the mixed tundra and forests found in Alaska today. All of this big game attracted the first humans, who followed in the footsteps of the megafauna across the land bridge. There is clear evidence that by about 14,000 years ago the first people had arrived.

Left: This mesa in Alaska is believed to have acted as a prehistoric hunting lookout for the first people in the region.
Right: Scientists use Paleo-Indian spear points as evidence for the presence of human hunters.

FIRST BERINGIAN HUNTING LOOKOUT

In 1978 a remarkable discovery was made on a mesa (flat-topped hill), which rises 60 m (200 feet) out of the tundra in northern Alaska. Fluted spear points were found alongside charcoal from fire hearths, established by carbon dating to be around 13,600 years old. This is one of the earliest pieces of confirmed evidence of humans living in Beringia. From the top of the mesa there is a 360-degree view of the surrounding area. With no sign of animals having been butchered here, it seems likely that hunters would have used the site as a hunting lookout, working on their tools and spear points while they waited by the fire for suitable game to appear. The site is only a few kilometres from the mountains where they would no doubt have retreated during winter, and it may well have been used all year round. But how did these first people live?

RADIOCARBON DATING

This is the technique used by scientists to date animal remains. Radiocarbon dating relies on the fact that cosmic rays continually transform small amounts of nitrogen in the atmosphere into a radioactive form of carbon: carbon-14.

Carbon-14 is unstable and decays to nitrogen-14 at a known rate. The half-life of carbon-14 is about 5730 years – that is how long it takes for half the atoms in any amount of carbon 14 to decay. Green plants assimilate carbon-14 during photosynthesis. Therefore all living plants are slightly radioactive, as are all living things, because they all feed indirectly or directly on green plants. Once an organism dies, its radioactivity decreases because it is no longer absorbing carbon-14. After 5730 years the remains of an organism will contain only one-half as much carbon-14 as the living organism did; after 11,460 years only one-quarter as much, and so on. By comparing the radioactivity of organic remains with that of living organisms, it is possible to tell how long ago a creature died.

Unfortunately the amount of carbon-14 can vary from year to year and so radiocarbon years are not exactly the same as calendar years. They can vary by as much as 15 per cent. 10,000 radiocarbon years correspond to 11,350 calendar years; 11,500 radiocarbon years to 13,350 calendar years. In this book we only use calendar years.

ALASKA'S FIRST PEOPLE

Even some 14,000 years ago, the first Americans were sophisticated hunters and gatherers, with the resourcefulness to survive some of the harshest environments on Earth. Small groups of people could now communicate with each other, sharing ideas and technology. Their accomplishments included tailored clothing; the control of fire; the ability to process, store and preserve food; cooperative hunting techniques; and fishing.

People probably carried leather bags containing a variety of tools and weapons with different uses. Perhaps the most important tool of all in this cold climate was also one of the simplest pieces of technology ever invented. The eyed needle enabled these people to make waterproof, warm clothing to keep out the Arctic cold. It has even been suggested that people would not have been able to enter North America via the windswept land bridge until they had invented the needle.

Although the first people would have made leather clothes and bags, these were quick to decay, and they only really left their mark in the region with the stone spear points, butchering tools and broken animal bones they left behind. The lack of evidence means there is still plenty of scope for controversy over the identity

Left: The first Americans entered a land of new opportunities. **Centre:** The final reconstruction of Blue Babe, the prehistoric bison mummy, now resides in the University of Alaska Museum, Fairbanks. **Right:** Palaeontologist Dale Guthrie excavating Blue Babe from a gold-mine.

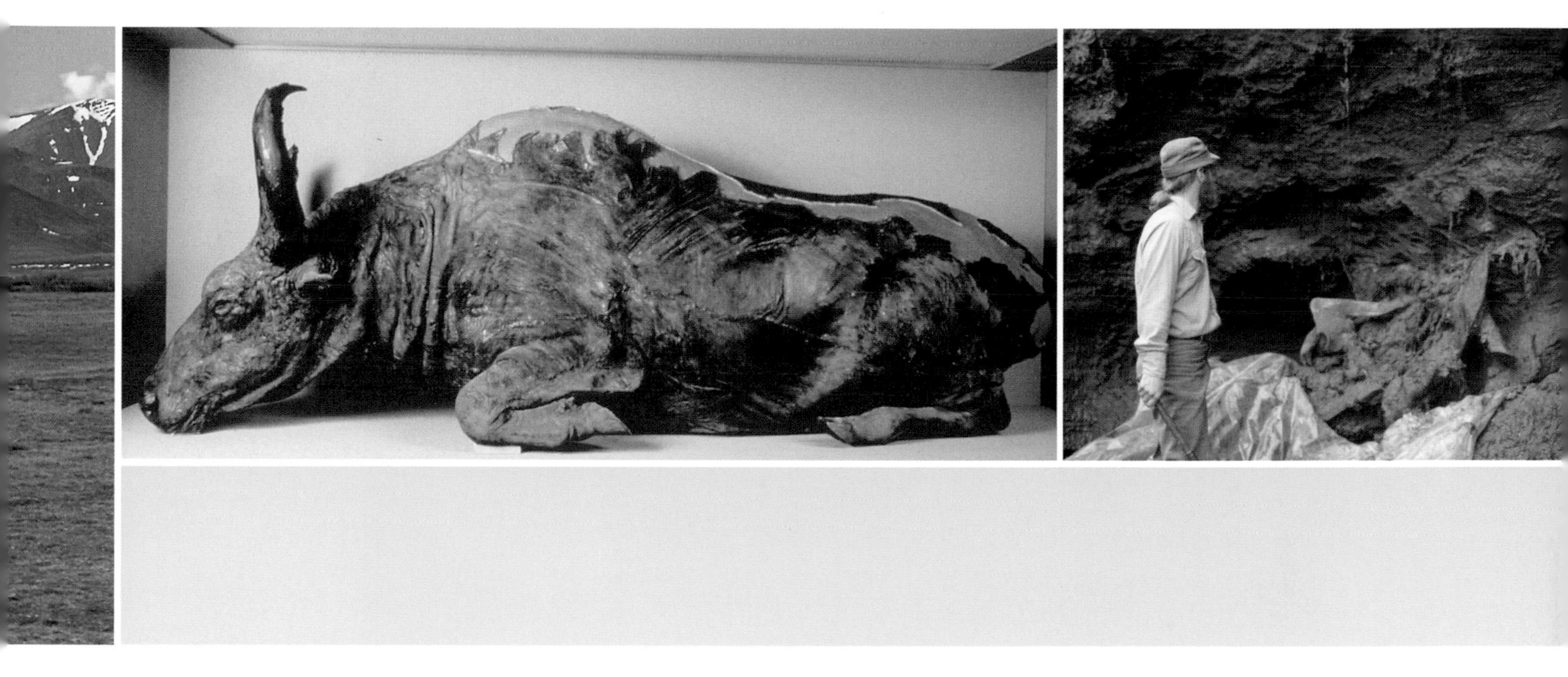

and movements of these first Americans. Whatever the arguments, it is now believed that several waves of colonization took place at different times.

During the past few years, a number of early human sites have been found. The Broken Mammoth site, discovered in 1990, overlooks the Tanana River, south of Fairbanks in Alaska. It is believed to be a Paleo-Indian hunting camp due to the presence of firehearth rocks and human artefacts dated at around 13,800 years old. These include spear points, atlatls (spear throwers) decorated with red ochre, and other tools made of stone and mammoth ivory.

BLUE BABE AND THE BIG CAT

People were not the only hunters in Beringia. Today the largest land predators in Alaska are grizzly bears and wolves, but during the Ice Age there were others. Evidence for the existence of an unexpected carnivore comes from the forensic examination of a 35,000-year-old mummified steppe bison.

In the summer of 1979 a placer miner discovered the hooves of a large animal protruding from a bank of muck just north of Fairbanks. Scientists supervised

African lions cooperate to kill large animals such as wildebeest. One grips the prey from behind with its claws while the other delivers a suffocating bite to the muzzle. Scenes like this would have been played out often in Beringia as American lions attacked bison.

the extraction of the body from the frozen earth to ensure that as much of it was preserved as possible. The chemical reaction of compounds on the hide when exposed to air turned parts of the body blue, so the male bison mummy was named Blue Babe after a giant ox that belonged to Paul Bunyan, the mythical hero of the northern forests.

The head and chest of Blue Babe were intact, but the hide on his back was torn, and much of the meat on the back and rump had been eaten. Perhaps he had simply died of old age or starvation? Annual growth rings in the horns and teeth suggested that he was only eight or nine years old when he died, and a thick layer of fat just under the skin indicated that he was in good health, so death by natural causes seemed unlikely. But if he was killed by a predator, who was the most likely suspect?

As the silt was cleaned away, long scratch marks were revealed in the bison's hide in parallel groups of three or four. This presented a mystery. Although wolves hunt bison, they do so by harrying them in a long drawn-out war of attrition. They do not use their claws but nip and bite with their teeth, giving chase till their quarry is exhausted. Grizzly bears have longer claws than wolves, but they are blunt because they walk on them and dig with them. Nor do they normally hunt anything larger than caribou.

The long scratch marks are much more like those made by a cat. The hide contained more clues: puncture wounds of large canine teeth about 8.5 cm (3.3 inches) apart. These must have belonged to a big cat, much bigger than the largest present-day North American cat, the cougar. There were further clues underneath the facial skin: more puncture wounds and evidence of blood clotting.

So, the prime suspect was a big cat that had jumped on the back of its prey, scratching and grabbing with its long claws and biting into the muzzle of its victim. If we move to Africa and substitute the American bison with its similar-sized counterpart the African buffalo, we discover that the buffalo's major predator is the lion. The African lion's hunting technique involves jumping onto the back of its victim, holding on with its long claws and finally clamping its jaws over the muzzle to asphyxiate its prey. The conclusive proof of the involvement of the King of the Beasts in the death of Blue Babe was the discovery of a tiny fragment of carnassial tooth in the bison's flesh. The enamel of the tooth fragment was particularly thick, as is found only in the lion family.

In fact, lion fossils have also been discovered in the Fairbanks area and it is known that the American lion was a significant predator of the Mammoth Steppe game. It should be noted that the cougar, also known as a mountain lion, is only a distant relative of the African and American lions.

We can now attempt to piece together the last day of Blue Babe's life. The presence of winter underfur in his coat puts the time of death as late autumn or winter. The attack was probably carried out by at least two lions because a fully grown, healthy bull would have been too difficult for a single animal to take on. The struggle to isolate him and bring him down was no doubt a long one but eventually Blue Babe collapsed onto his chest and into the position in which he would be discovered 35,000 years later. After the asphyxiating bite, the lions would have fed on his back and rump. They probably visited the carcass a number of times over the next couple of days to feed. But by then the body would have been frozen hard and difficult to gnaw: the chip of tooth left in the hide was testament to this. So, once the body was frozen solid, the lions probably left it for easier fare. Blue Babe remained frozen until he was covered over by a blanket of silt and loess in the spring thaw. Successive thaws created further layers until he was completely buried in frozen silt – the perfect medium for mummification.

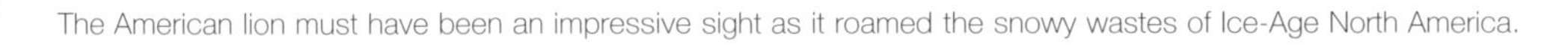

The American lion must have been an impressive sight as it roamed the snowy wastes of Ice-Age North America.

THE AMERICAN LION

Blue Babe's killer, the American lion, was found throughout the continent during the last Ice Age. Lions evolved in Africa around 700,000 years ago; from there they spread to Asia and Europe, as well as North and South America. American lions are believed to have been up to 25 per cent larger than their living African relatives. They were well adapted to the range of climates and environments found in North America – from the Arctic steppe, through forests and grassland, to the southern deserts. They became extinct around 10,000 years ago, probably because of the disappearance of many of their prey.

Socially, American lions appear to have behaved somewhat differently to typical African lions. In some populations of African lions, the animals live in pairs rather than as members of the well-known large prides of the savannah. This occurs when a lack of prey at certain times of the year prevents the formation of larger groups. Many believe that the distribution of game in Beringia would have favoured lions operating in pairs or small groups. Male African lions living in small groups have smaller manes, perhaps because the intimidation of rivals is less crucial than in the larger groups. So perhaps American lions also had smaller manes. Evidence from ancient European cave paintings certainly suggests that northern lions may have had smaller manes than modern African lions.

The study of Blue Babe shows that Beringian lions would have hunted bison. Two lions would normally have been required to bring down an adult bull, but single lions may well have hunted females and calves. In addition, there would have been plenty of horses, caribou and muskoxen for them to feed on. Lion fossil specimens have been found in the Yukon showing healed breaks on the jaw, where the animals had probably been kicked by horses or other hoofed prey. Finally, African lions have been known to attack baby elephants, and their American cousins would no doubt have taken advantage of an unguarded baby mammoth but, given the protective behaviour of elephant families, this would probably have been a rare event.

While the American lion was one of the largest of Beringia's predators, it is likely that even it feared one animal. This was another creature whose existence has been revealed by the hoses of placer miners: the giant short-faced bear.

GIANT SHORT-FACED BEAR

KEY FACTS

Common name: Giant short-faced bear

Scientific name: *Arctodos simus*

Size: Over 3 m (10 feet) tall on hind legs

Weight: 700 kg (1540 pounds)

Diet: Carnivorous

Habitat: Steppes and woodland

The giant short-faced bear was 1.5 m (5 feet) tall at the shoulder and rose to an impressive 3 m (10 feet) when standing on its hind legs. This giant was taller than a polar bear and twice the weight of a grizzly, but with slimmer legs. The short muzzle gave it a more lion-like face than other bears. It had a relatively wide skull and very powerful jaws. Its closest living relatives are not the grizzlies, but the South American spectacled bears.

For some experts, the longer legs and powerful jaws of the short-faced bear indicate that it was a fast-moving predator – a kind of cheetah bear, which may even have prevented human colonization of North America until it had become extinct.

However, at 600–800 kg (1320–1760 pounds) it was probably too heavy to be a specialist predator. Its ability to accelerate and manoeuvre would have been limited, and its long legs may not have been strong enough to sustain a chase. Instead it seems that this animal was built for ranging widely across the landscape with minimal effort.

But if it was not a hunter, what did it eat? Most bears, such as black and grizzly bears, are omnivorous and feed on animal prey when they can get it, but also eat berries, tubers, roots and even grasses. The only exception to this is the polar bear – a marine specialist that preys on seals.

Fortunately the old saying 'You are what you eat' proves true in the study of fossils. Analysis of the bones of short-faced bears shows that they were exclusively meat eaters. And they were well adapted to this task. Their shortened jaws would have brought their crushing teeth closer to the back of the skull and so have increased their power. This bear appears to have been adapted for cracking large bones to extract the nutritious marrow. But if the giant short-faced bear did not hunt, how did it come by the meat and bones that made up its diet?

Left: A herd of muskoxen responds to a passing giant short-faced bear by huddling together with their horns turned towards the threat.
Right: On its hind legs, the giant short-faced bear was very intimidating. As the largest North American carnivore it could take its pick of the carcasses killed by other carnivores.

Ice-Age distribution of giant short-faced bear

An important clue comes from the animal's wide snout, which would have provided enlarged nasal openings and probably heightened its sense of smell. So this animal was a bone-cracking carnivore with an excellent sense of smell, adapted for long-distance travel. In other words, a long-range scavenger. And, with so much big game around, Ice-Age America was a great time to be one. The abundance of predators such as American lions, scimitar-toothed cats and wolves meant that, for the short-faced bear, there were always carcasses to be taken from smaller animals. Indeed the short-faced bear may have evolved to such a great size just to intimidate the other carnivores.

While basically a scavenger, the bear would no doubt have been an occasional predator. Some grizzly bears hunt elk or moose calves, but only during the brief three or four weeks in which the calves are too small to outrun them. Other grizzlies occasionally test herds of caribou by running into them to look for sick or young animals. It is unlikely that a short-faced bear would have turned its highly developed nose up at a similarly easy meal.

COPING WITH THE COLD

During the Ice Age (as today) the most notable feature of Alaska and the Yukon was its harsh climate. Winters were long and dark, so how did animals survive? Larger animals cannot avoid the cold and rely partly on feeding up in the summer and autumn to create fat reserves. They also have specialized coats consisting of long guard hairs (some of which may be hollow for extra warmth) and short, dense underfur. Saiga antelopes, Dall's sheep, muskoxen, bison and woolly mammoths all conform to this pattern. However, smaller animals would have had to use some form of shelter to survive the worst of the weather.

The Arctic ground squirrel relies on its long hibernation, which can be eight months in parts of its range, to get through the winter. Alone in its hibernation chamber, it drops its body temperature to -2°C (28°F) to reduce energy loss during this extraordinarily long period of time. Mummified ground squirrels have been found in hibernation burrows having never woken up. One of the oldest is a 47,000-year-old specimen found near Dawson in the Yukon. The squirrels were as numerous then as they are today and were a vital part of the Beringian food chain.

Pikas are small and round, with short legs and little ears to reduce heat loss. They live alone on the bleakest rocky slopes at the tops of the high nunataks, which they claim by sitting on a prominent rock and whistling. Remarkably they do not hibernate, and they make the most of the near 24-hour daylight during the brief four-month summer to prepare for winter. They collect grass and flowers and stack them in large hay piles up to 50 cm (1.6 feet) deep, sheltered beneath a particularly large slab of rock. Individuals in neighbouring territories will steal from each other's hay piles. In the winter the pikas remain active within the spaces between the jumble of rocks, oblivious to the wind and cold outside, feeding on their refrigerated larder. Most of the world's pikas are Asian, and so the two North American species must have migrated across the Bering land bridge with the larger animals. Just how these diminutive creatures travel across the icy, exposed slopes from one nunatak to another is still a mystery. Their existence on the remotest peaks is a testament to their powers of endurance. There is evidence that today's nunataks existed during the last glaciation, and so the pikas may well have lived on them throughout the Ice Age right up to the present day. Perhaps the tiny pika deserves to be an Ice-Age icon just as much as the woolly mammoth.

Left: The Arctic ground squirrel during its eight-month sleep. **Centre:** The Arctic ground squirrel makes the most of the brief summer by feeding constantly. **Right:** The tiny collared pika works hard all summer to collect enough food to see it through the long winter months.

THE END OF BERINGIA

It is impossible to underestimate the achievement of the first people who endured the hardships that nature imposed on them as they crossed this unforgiving land. Around 11–12,000 years ago the land bridge to Asia was finally covered over by the sea, and for the people and animals that had come across from Siberia there could be no return. From now on they were American and they had to make the most of what the new continent had to offer.

For these people there was a potential hunting paradise to the south – a vast array of big game, most of which had never seen humans before. As the Ice Age came to an end, the retreat of the glaciers opened up two possible routes that allowed people to leave the ice-locked Beringia and explore the south of the continent. The first route was an ice-free corridor east of the Rocky Mountains, which led to the central plains; the second route was along the milder coast of the north-west and it is here that we continue our journey in the next chapter.

Two female adult woolly mammoths keep a wary eye on a giant short-faced bear as it sniffs the air in search of a meal. In the distance wild horses graze the steppes of Beringia.

THE NORTHWEST COAST

Left: Ferns, mosses and lichens thrive in the humidity of the temperate rainforest. **Above centre:** Snags – free-standing dead trees in a forest – support a rich variety of wildlife. **Below centre:** Grizzly bears are the largest carnivores in the northwest. **Right:** Black-tailed deer inhabit temperate rainforests, though it is rare to catch a glimpse of them.

Tidewater glaciers are modern-day reminders of the Ice-Age past.

ANCIENT FORESTS AND RIVERS OF ICE

The Pacific coastal region from Oregon to southeast Alaska is home to a true wilderness that evokes a prehistoric landscape untouched by time. This is the temperate rainforest zone. Extending over 3200 km (2000 miles) from north to south, the mainly Sitka spruce-dominated forests found here account for half of the world's temperate rainforests. Moisture is the key ingredient. An average of 3.6–4.3 m (12–14 feet) of rain falls a year and wispy low blankets of fog frequently wrap themselves around the forests, softening their edges and making them resemble a surreal watercolour of deep greens, blacks and greys.

TEMPERATE RAINFORESTS

The temperate rainforests found along this stretch of coast contain some of the tallest trees in North America. Sitka spruce can grow to a height of over 90 m (300 feet) and live for up to 700 years. Even when dead, these trees play a vital role in the ecology of the rainforest by supporting a variety of animals and plants. The rainforests absorb moisture and the lush vegetation that results includes thousands of different plant forms that grow on and over the trees. Plants that use other plants for support are called epiphytes. They are like rent-free tenants of a giant apartment block. In temperate rainforests, epiphytes such as ferns and mosses cake the branches of trees. Over time the dead remains of these squatters can build up a layer of soil 30 cm (12 inches) deep on the limb of a tree.

Travelling inland from the coast, the temperate rainforests give way to drier lowland forests. Sitka spruce is no longer as prevalent but giants such as Douglas fir and western hemlock grow. The coastal mountains soon rob the rising sea air of its moisture and, with increasing altitude, the forests thin until the last trees cling to the subalpine slopes. Snow is the main form of moisture here and the cold winters stunt forest growth.

Like their tropical counterparts, temperate rainforests are home to a rich variety of animals. The largest inhabitant is the black bear, which, though a predator, spends much of its time grubbing around on the forest floor for berries, insects and fungi. Sitka black-tailed deer – a type of mule deer – favour old-growth forests, where they forage on berries, shrubs and trees. Their hearing is extremely acute and any glimpses of them are usually restricted to the sight of their bobbing black tail as they flee into the depths of the forest.

The primeval feel of old-growth temperate rainforests makes it difficult to imagine a time when they did not exist. But although they are often described as 'ancient forests' they are in fact comparatively recent, geologically speaking, at little more than 5000 years old. Trees recolonized new land exposed by the retreating ice sheets 14,000 years ago, but it took many thousands of years for the forests to mature and develop into what we see today. This process of recovery is still visible in the modern-day landscape.

Southeast Alaska (also known as the Alaskan panhandle) encompasses the narrow strip of land that joins Alaska to Canada's British Columbia, as well as the thousands of forested islands alongside it. This is a place of stunning contrast with great swathes of temperate rainforest growing next to blue-iced glaciers,

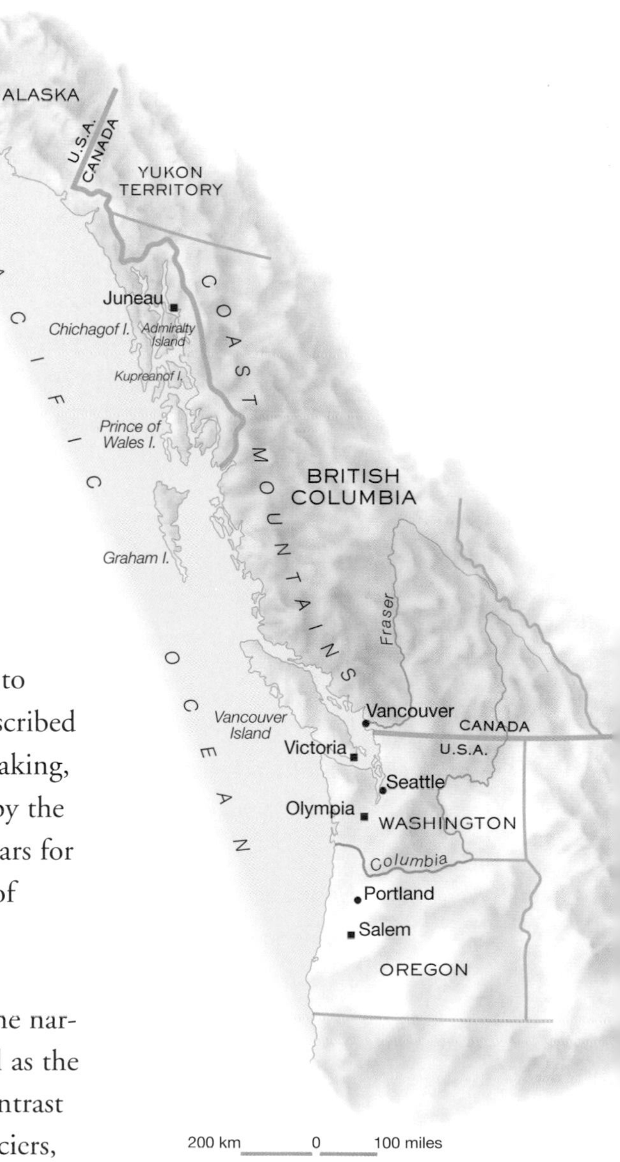

Left: The upland forests of Misty Fjord National Monument, southeast Alaska. **Right:** From the air, forests can be seen recovering ground around the retreating snout of a glacier.

which plunge into the water of long, narrow fjords. Today southeast Alaska has 70 glaciers covering 10 per cent of the land, a tiny fraction of the ice that once blanketed this area during the peak of the last Ice Age. The glaciers we see today are fragmentary remains – most of them are in retreat and the new land around them is being recolonized by plants. This is particularly visible from the air, where young trees can be seen just a few metres from the receding glacier front.

So with a little imagination it is possible to picture the landscape that greeted the first people arriving in this region as the great ice sheets were melting towards the close of the last Ice Age. Clues to the past can still be found in the living landscape of the present-day northwest, often buried for thousands of years under ice, under forests or occasionally under towns. Fortunately the past has a habit of resurfacing from time to time and, when it does, we get a privileged glimpse of another era.

SNAGS

For many years scientists believed that the standing dead trees, or snags, of coastal forests were an indicator of poor health in the overall ecosystem. But these eerie wooden skeletons are now known to contribute to the ongoing diversity of the forests. They provide a home for mammals such as tree voles and flying squirrels and are important nest sites for birds such as woodpeckers and owls, including the rare northern spotted owl.

Snags may result from the natural death of a tree, but they are often caused by insect activity. Douglas fir is a dominant tree in the lowland temperate rainforests of the northwest. The Douglas fir beetle takes advantage of trees weakened by drought or damage. Every spring, female beetles lay eggs in the wood. They are also the carriers of a fungus, which spreads throughout the wood, blocking the water-carrying channels that are essential in keeping the tree alive. Locked in a slow stranglehold, the tree slowly dies through lack of water and nutrients and yet it can remain standing for many years before toppling to the forest floor.

SMALL TOWN, BIG BEASTS

In the small town of Woodburn, near Portland in Oregon, a field next to a local high school has become the focus of attention for one of the biggest archaeological digs in the Pacific northwest. As with so many significant finds, the site was discovered by accident – in this case during the routine cutting of a sewer pipe trench in 1987. After several summers of ongoing excavations, the fossilized remains of more than 20 species of mammals have been recovered and the digs have helped to create a convincing picture of the northwest towards the end of the last Ice Age. Elephant-like creatures such as mastodons and Columbian mammoths lived alongside a giant form of ground sloth; horses and bison mingled with bears and wolves.

The remains of all of these animals were buried in bog-like conditions over 10,000 years ago. The beauty of Woodburn is the clear stratification of the sediments, with each layer taking us further back in time. It is rare to find sites with such clear layering, and the dating of the fossils is made much easier as a result. Fossils in the same layer will have come from animals that died at a similar time.

The Woodburn finds tell us that the landscape of the Pacific northwest was fairly mixed at the end of the last Ice Age – browsers such as mastodons fed on trees and shrubs, while grazers such as mammoths, horses and bison fed on grass. The dense forests found in the area today were then confined to their Ice-Age refuge further south, but trees were, nonetheless, slowly making a recovery. The discovery of a type of giant ground sloth (see Chapter 5, p. 151) has added to the richness of the finds. In addition, Woodburn has yielded a very rare fossil – the remains of a baby ground sloth, which may either have died just after being born or have still been in the womb when its mother died.

Apart from the many different mammal finds, the Woodburn sediments also contained the tiny bones of birds, including those of ducks. This suggests the presence of inland lakes or bogs. However, the most impressive bones belong to a large prehistoric bird. With a huge wingspan of 4.9 m (16 feet), this is one of the first Pacific northwest recordings of the now extinct teratorn (see p. 59).

Even the smallest finds can help us to reconstruct the past. Thousands of iridescent beetles have been recovered from Woodburn; many may be previously

Dry Falls in Washington State is a geological reminder of huge Ice-Age floods.

LAKE MISSOULA FLOODS

The waxing and waning of the ice sheets exerted a great influence over the North American continent. The Pacific northwest in particular bears the scars of the catastrophic consequences of melting ice.

Just over 12,000 years ago this area felt the impact of one of the world's greatest ever floods. In the state of Idaho a retreating glacier damned the Clark Fork River, leading to the creation of a giant glacial lake – Lake Missoula. At 7700 sq km (2970 sq miles) and 610 m (2000 feet) deep, this lake was more like an inland sea. The ice dam could only hold back this huge volume of water for so long; it eventually collapsed and water exploded out of the lake. In 48 hours the lake emptied, with a huge tower of water travelling westwards across Idaho and Oregon at 105 km/h (65 mph). The water scoured the soil and penetrated deep into the bedrock, creating spectacular canyons and giant ripple marks, still visible today. These floods were equivalent to 10 times the outflow of all the world's rivers today. In their wake they left a scarred landscape with huge gravel deposits. The remarkable thing is that Lake Missoula flooded not once, but up to 100 times. The floods must have wiped out millions of animals and may have killed some of the first people in this part of the continent.

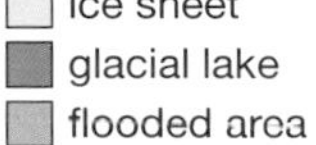

undiscovered species. Bark-boring beetles indicate that trees were present. Carrion beetles were clearly making use of rotting meat, sharing carcasses with scavenging birds and mammals. Although some of the insect species recovered from the sediments no longer live in this region, many of the insects around Woodburn at the end of the Ice Age can still be found in the Pacific northwest. It seems likely that temperatures were similar to those today, but the retreating ice sheets continued to hold sway over the climatic conditions. Rain patterns were in a state of flux and the increased precipitation that would eventually support the recovery of the forests had yet to dominate the climate around Woodburn.

Recent discoveries indicate that the first people to inhabit the region may have witnessed sabre-toothed cats hunting horses and teratorns scavenging bloated mammoth carcasses. Within the sediments, tiny basalt and chert flakes – by-products of human stone-tool production – have been dated at over 12,000 years old, some of the most ancient finds in this area.

However, the find of greatest significance has been the discovery of human hairs, complete with roots. The exact dating of the hairs is controversial, but radiocarbon tests suggest a rough age of 12,000 years. If the dates are correct then this will be one of the earliest examples of actual human remains to be found on the North American continent. Direct evidence of people living alongside extinct Ice-Age animals in North America is not that common and fossil finds that inextricably link early North Americans with the hunting of these beasts are bound to generate considerable interest.

MASTODON MANIA

In 1977 the small town of Sequim on the Olympic Peninsula in Washington State became the focus of a huge amount press attention. The discovery of a mastodon in a farmer's field in the shadow of the Olympic Mountains became more significant when what appeared to be a human-made bone spear point was found within one of the animal's ribs. The Manis mastodon, as it came to be known, was named after the man who made the discovery – Emanuel Manis. While excavating a pond with a mechanical digger, Manis pulled out what at first appeared to be a log. It soon became apparent that he had stumbled across

A complete mastodon skeleton from Illinois. This part of North America has yielded some of the best mastodon finds.

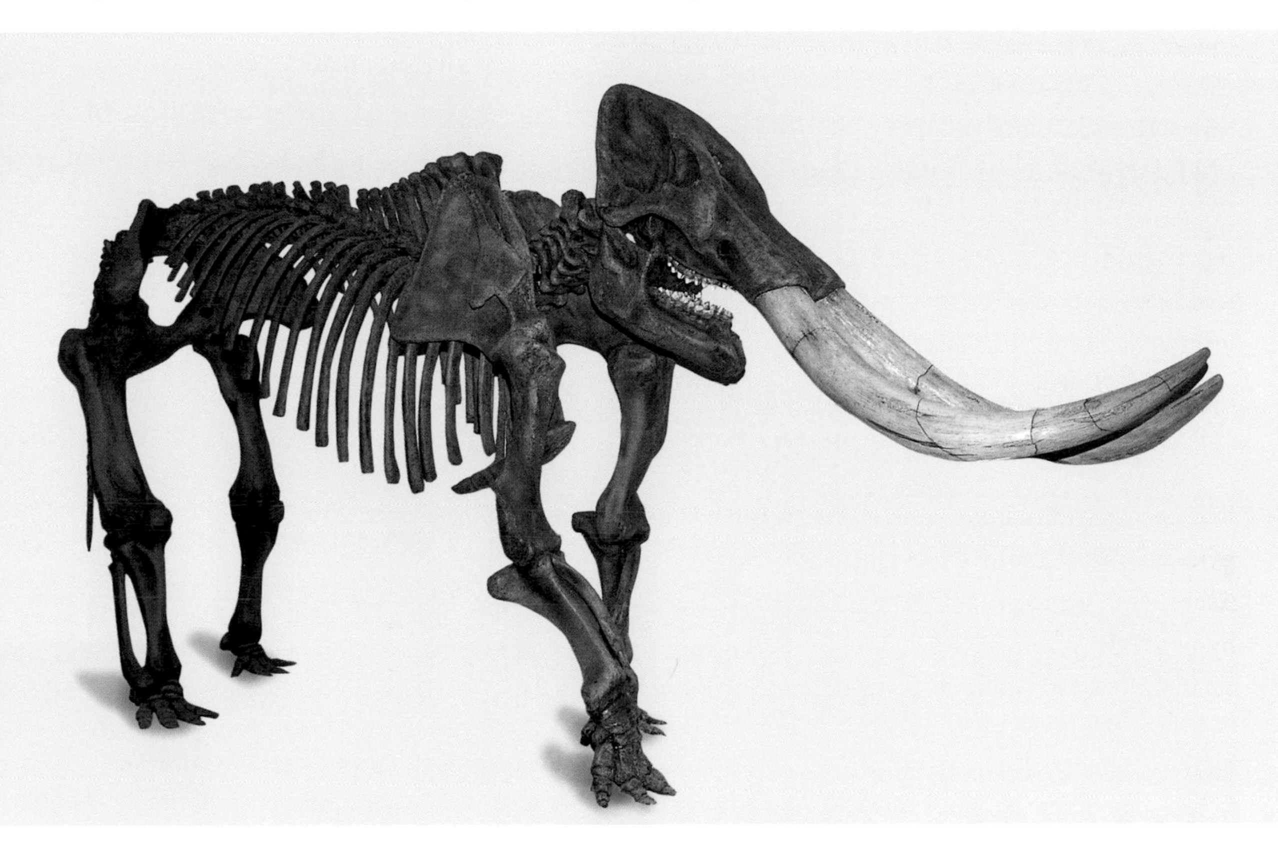

an elephant tusk. Within a week a major archaeological dig was under way, and during the first two hours of digging the rib fragment with the spear-shaped bone intrusion was discovered.

The Manis mastodon fossil is dated at around 14,000 years old. Weighing in at a hefty 6800 kg (6.7 tons), these fossilized bones were obviously from an impressive bull. The left-hand side of the skeleton was more or less intact, with bones from the right-hand side scattered around the area. The apparent bone spear point in the mastodon's rib implicated people in its death, but closer examination revealed that new bone had grown around the damaged rib. So, if humans had indeed attacked the mastodon, they didn't kill it. Instead the old bull recovered from his wound and lived on.

The excavations took place over several summers, and thousands of people flocked to the Manis farm to find out more about this ancient beast. Many of them would not have heard of a mastodon before and would have been asking the question: what exactly is it?

MASTODON

KEY FACTS
Common name: American mastodon
Scientific name: *Mammut americanum*
Size: 2.4–3 m (8–10 feet) tall at the shoulder and 4.5 m (15 feet) long
Weight: 3600–5500 kg (3.5–5.4 tons)
Diet: Vegetarian – browsing on trees (mainly coniferous), shrubs and swamp plants
Habitat: Forests, open parkland and swamps

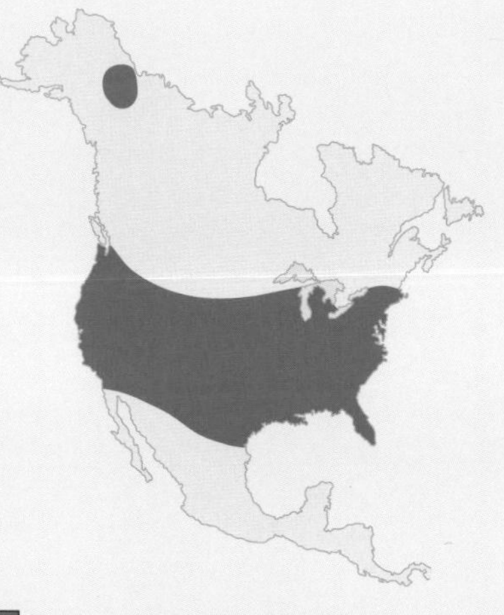

Ice-Age distribution of mastodon

If the extinct megafauna of North America were to be given celebrity status, then the American mastodon would surely fall into the B-list category. Mammoths have always received more publicity and attention than mastodons. Maybe this is due to their size: at 2.75–3.4 m (9–11 feet) tall, woolly mammoths were comparable in height to modern elephants, while mastodons were slightly shorter at 2.4–3 m (8–10 feet). But despite being smaller, mastodons were stockier, with thicker limb bones, and they could have weighed as much as a fully grown mammoth at 3600–5500 kg (3.5–5.4 tons).

Mastodons belong to an ancient family called the Mammutidae, which originated in North Africa some 30–35 million years ago . This makes the mastodons an older group of elephantine animal than the mammoths, which did not appear until around 10 million years ago. In this sense the mastodon is one of the truly prehistoric creatures of North America. As with much of the megafauna in North America, an earlier form of Pleistocene mastodon entered the continent via the Bering land bridge around 15 million years ago. This ancestral form ultimately gave rise to only one Pleistocene species of mastodon in North America. Its closest relative, the Borson's mastodon, lived in Europe about 3 million years ago.

Mastodons would have lived through the many glaciations that covered the North American continent during the past 2 million years. At the end of the last glaciation, they were widespread across the whole continent. However, most fossil finds suggest that mastodons were most common on the eastern seaboard and around the Great Lakes.

So far there have been over 200 finds of mastodon fossils across the whole of North America. Less is known about the anatomy of mastodons than mammoths because few have been preserved in frozen form. Mastodons had straighter tusks than mammoths and were probably less hairy than woolly mammoths. One of the key distinguishing features between mammoths and mastodons was their teeth.

Mastodon cheek teeth consisted of pairs of conical cusps, which resembled a miniature mountain-like landscape. Mammoth teeth, on the other hand, were more like

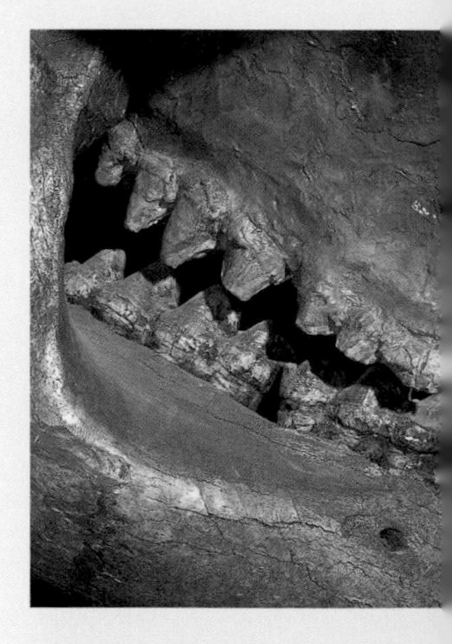

Centre above: Mastodon teeth. **Centre below:** Mammoth teeth. **Right:** An adult bull mastodon.

elephant teeth, with rows of enamel ridges serrated to form a washboard-like surface. The teeth place the mastodon as a primitive and distant relative of mammoths and elephants. The very name mastodon means 'breast tooth' – a reference to the nipple-shaped conical cusps.

Although mastodons often lived in the same regions as mammoths, their teeth shape suggests a different diet and habitat. The high-ridge conical cusps created a grinding surface that could be used to wear down hard material like branches and conifer needles. Indeed some preserved mastodon teeth have been found with small amounts of conifer twigs between the cusps.

Like elephants, the mastodon could use its tusks and trunk to snap off branches of trees. Some mastodon finds show the left tusk to be longer than the right, suggesting a left-tusk bias in the same way that most humans have a right-handed bias.

Swamp plants and mosses have also been found in association with mastodon teeth. Like modern-day moose, mastodons probably gained extra minerals from eating water plants of swamps and ponds. A giant form of moose, the stag moose, often lived alongside mastodons but it too went extinct at the end of the last Ice Age. Today, moose occupy habitats similar to those that mastodons and stag moose once lived in.

The social behaviour of mastodons is still a controversial issue. Many of the fossil finds have been of solitary animals, leading some to suggest that mastodons did not live in herds, or at least not in large numbers like elephants. Both mastodons and mammoths were heavier than elephants and may have had a longer gestation period. Mastodon young are thought to have taken around 10 years to reach sexual maturity, thus requiring a longer period of maternal protection than elephants. Given such a long period, it would not be surprising if mastodons lived in matriarchal groups. The long-term investment in raising offspring is a risky strategy that benefits from protective herding.

Caribou were widespread in the northwest during the last Ice Age, which suggests that there was an open, tundra-like landscape.

CARIBOU, CACTI AND OLD BONES

If humans didn't kill the Manis mastodon, then how exactly did it die? To find out we must learn more about its surroundings and the animals that shared its environment. The majority of fossils scattered around the Manis mastodon come from caribou and bison. This suggests that the mastodon was not in its preferred habitat when it died. Mastodons are known to browse on trees, particularly conifers such as spruce. Caribou, on the other hand, are a tundra-loving species, spending their lives on the move as they migrate between their calving grounds and their wintering grounds. Today they are found in the far north of the continent, in the northern latitudes of Canada and Alaska. During the last

Ice Age the glaciers pushed them southwards. Bison, meanwhile, are grass-loving animals familiar to us as the symbol of the Wild West. Today bison and caribou live their lives thousands of miles apart in very distinct habitats. The presence of their fossils in the same place suggests that the clearly differentiated habitats existing today were not present at the end of the last Ice Age.

This is further supported by pollen samples taken at the same depth as the fossil finds. Remarkably no tree pollen has been found, a little odd since mastodons were browsers. However, shrubs like willow were present and an unusual pollen type also made an appearance – cacti. It is hard to imagine cacti in a region that is now home to some of the world's best temperate rainforests, though cacti are not always found in deserts and even today prickly pears occur in Washington State and British Columbia. Their clear presence in the pollen record from 14,000 years ago tells us that some areas must have been dry. Despite being in retreat, the enormous ice sheets were still able to exert great influence over the region's climate. The rainfall that characterizes the Pacific northwest today was pushed much further south. Trees were yet to recover on the new land exposed by the retreating glaciers not because the environment was too cold but rather because it was too dry. That said, shallow ponds and lakes did punctuate the landscape, some of which were seasonal and would have dried out during warm summers. Around the ponds and lakes, sedges and grasses grew, mixed with shrubby willow and sagebrush, as well as the occasional cactus.

What was a browser like the mastodon doing in a treeless environment? It is possible that some trees were present, but too few to register in the pollen record. Perhaps the mastodon had strayed far from the nearest trees and was lost. However, other mastodon remains have been found near the site, suggesting that the Manis mastodon was not an isolated individual. If, as the pollen record shows, the climate was drier 14,000 years ago, a pond or small lake would have been a focal point for many animals. Recent evidence in Florida suggests that mastodons migrated over long distances (see Chapter 5, p. 154). Whether the Manis mastodon was lost or had deliberately travelled some distance, there is no doubt that a pond or lake would have been a big draw. Watering holes are also attractive to predators as they provide a concentrated source of potential prey. Predatory animals appear less often in the fossil record and none were found at the Manis mastodon site. There is, however, plenty of evidence elsewhere on the continent that points to one animal being the most successful predator of mammoths and mastodons – the scimitar-toothed cat.

SCIMITAR-TOOTHED CAT

KEY FACTS
Common name: American scimitar-toothed cat
Scientific name: *Homotherium serum*
Size: 1.1 m (3.6 feet) tall at the shoulder
Weight: 100–200 kg (220–440 pounds)
Diet: Carnivorous – known to attack the young of mammoths and mastodons
Habitat: Mixed, from grassland to open forests

Like the mastodon with the mammoth, the American scimitar-toothed cat (or scimitar) has not enjoyed as many hours in the glare of publicity as its meatier relative, the sabre-toothed cat. This again may be due to size – scimitars were well endowed with large canine teeth, but they were not as big as those belonging to sabre-tooths. Yet it is the scimitar that appears to have been the strongest contender for the title of 'mammoth killer'.

The American scimitar probably evolved from a cat that ranged throughout the northern hemisphere some 2 million years ago. Despite its slender limbs, the scimitar was the size of a lion. Although found across the whole of North America, the scimitar seems to have been a rarer beast than the sabre-tooth. Several key features distinguished these two cats. Scimitars and sabre-tooths both had enlarged canines but the scimitar's were shorter and more slender than the sabre-tooth's. As the name suggests, the teeth of scimitars were finely serrated and acted as powerful slicing tools.

When first discovered, the scimitar was thought to be a type of bear. Indeed, it is hard to picture a scimitar without making comparisons to modern-day animals. It has been likened to bears, hyenas and cheetahs, and it shares features with all of these animals.

The front half of the scimitar suggests that it was a fast animal capable of carrying heavy objects. Elongated forelimbs ended in partially covered claws – a characteristic shared with modern cheetahs – which may have given the scimitar an enhanced grip during a sprint. In addition to its elongated canines, the scimitar also possessed relatively large incisors, suggesting that it was capable of lifting heavy weights and tearing at thick hides. Strong jaw muscles would have given the cat a very powerful biting grip. Just as tigers today can be seen carrying whole kills using their strong jaws and long legs, so scimitars may have done the same.

The back half of the scimitar is a different matter. The hindlimbs were shorter than the forelimbs and the scimitar probably had a sloping back that would have made it look bear- or even hyena-like. Although it shares features with modern-day carnivores, the scimitar had a unique look and was probably built for speed. This is further suggested by its enlarged nasal passages, another feature that it shares with the sprinting cheetah. It is thought that the enlargement of these chambers facilitated greater oxygen intake, helping to fuel the release of energy from working muscles.

Evidence to support the theory that the scimitar was the greatest killer of mammoths and mastodons comes from a variety of fossil finds where cat remains have been found alongside juvenile mammoths. The largest such find was made at Friesenhahn Cave in Texas during the early twentieth century. The remains of some 300 to 400 juvenile mammoths were discovered alongside more than 30 adult scimitars and two kittens. This

Above: Young mammoths, straying from the family group, were prime targets for predatory scimitar-toothed cats. **Centre:** The fossilized skull of a scimitar. **Right:** A snarling scimitar reveals its large razor-like canines.

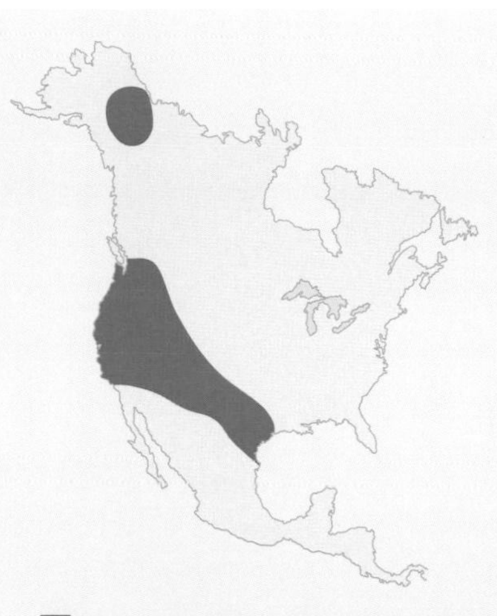

suggests that the cats may have been social in their behaviour. It is possible, however, that although cats used this cave they were not necessarily team players in their hunting strategy. The sheer number of mammoth and mastodon fossils reflects the scimitar's effectiveness in hunting the largest animals on the continent. Whether they hunted in pairs or on their own, scimitars would have used speed and ambush to kill their prey.

It has been argued that the ability of the scimitar to penetrate the matriarchal ring of mammoths and mastodons proves that the giant elephants of North America did not have such a strong group protection as their African relatives. Scimitars probably sought out the most vulnerable animals, whatever the level of protection. The majority of remains in Friesenhahn Cave come from mammoths around the age of two – just when they are beginning to separate from their mothers and are easier to attack. Using speed and its serrated teeth, a solitary scimitar could have ambushed and cut the throat of a juvenile mammoth and made its retreat long before the mother had time to react. The scimitar would then wait for the juvenile to die and the family to move on before claiming its meal. With its powerful jaws it would tear off great chunks of the carcass and carry its prize to the safety of a nearby cave, to eat it undisturbed.

Scimitars would have been attracted to any location where mastodons gathered but it is unlikely that they would have tackled a fully grown adult. To discover the real reason for the Manis mastodon's death we must investigate his bones. With 2 m (6.5 feet) long tusks, he was clearly very mature. His teeth showed signs of extreme wear, the ridge-like cusps having worn smooth. They were perilously close to falling out. African elephants lose their teeth at around 45 years of age and this tells us that the Manis mastodon was somewhere around that age. Some of his bone joints were swollen, perhaps from arthritis. An unusually heavy grass diet may have accelerated the wearing down of his teeth, but in the end it is likely that this magnificent beast simply died of old age. Whether he was alone or in a group of other mastodons we may never know.

Scavengers would have soon descended on the 6300 kg (6.2 tons) of available meat, a passing scimitar no doubt taking much of the carcass. The fossil evidence from Woodburn, just south of the Manis find, suggests other scavengers might have been present, including the teratorn.

Ultimately the mastodon attracted people. Many of the bison remains around the mastodon show spiral fractures, commonly associated with human butchering. Furthermore, the mastodon skull, though in tiny fragments, had been turned through 180 degrees so that it was facing backwards, and the back of the cranium had been smashed in. The scatter of bones, the damage to the skull and the discovery of human tools nearby indicate that people played a part in the butchering of the animal after its death.

They must have arrived soon after the animal died if they were butchering it for fresh meat. Perhaps they scared away some of the larger scavengers. They took what they could, but the mastodon had died on the edge of a pond or lake, and its left-hand side was buried underwater. What remained soon sank into the boggy bottom of the lake.

The presence of people on the Olympic Peninsula around 14,000 years ago continues to raise the question of where they came from. Perhaps they had travelled overland, crossing the Olympic Mountains. It has long been thought that the first people in North America spread to the coasts from the interior. However, in recent years, the Pacific northwest has become the focus of research that suggests a route previously thought impossible.

Bison were present on the Olympic Peninsula at the end of the last Ice Age.

TERATORNS

Teratorns were the giants of the Ice-Age bird world. Much of the evidence for them comes from the tar pits of La Brea in Los Angeles (see Chapter 3, p. 95). Merriam's teratorn had a wingspan of 3.7 m (12 feet) – wider than any bird alive today. Condors reach, on average, 2.7 m (9 feet) and albatrosses 3 m (10 feet). But even Merriam's teratorn was dwarfed by its relative, the Incredible Teratorn, which had a wingspan of 4.9 m (16 feet).

Teratorns were related to New World vultures, which belong to the same family group as storks and herons. Debate continues as to whether the teratorn was a scavenger, like its relative the vulture, or an active hunter. Reconstructions of the beak give the bird an almost eagle-like appearance. Teratorns may have hunted small mammals and eaten them whole but their relatively puny talons (compared to modern-day eagles) probably eliminated them as active aerial hunters that grabbed their prey from above. If teratorns were active predators it is likely that they worked on the ground. Like most animals, teratorns would certainly have scavenged. More than 100 of these birds have been found in the La Brea tar pits and there is little doubt that they gathered to enjoy the easy pickings of large mammals caught in the sticky tar of Pleistocene Los Angeles.

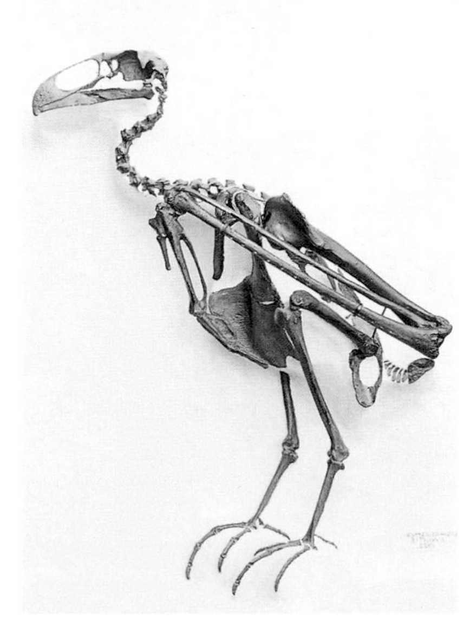

Left: A brown bear, or grizzly, from Admiralty Island in southeast Alaska. **Right:** Ringed seals were present in the waters of southeast Alaska at the end of the last Ice Age.

BEAR BONES AND GRIZZLY GENES

Geologists have long been convinced that the Pacific northwest was completely covered in ice during the peak of the last Ice Age. Glaciers spread out onto the continental shelf making it inhospitable for living creatures, so the theory went. However, this picture of total ice cover has been seriously challenged by recent finds in digs carried out between 1994 and 2000 in southeast Alaska.

Prince of Wales Island is the largest island in this region and the third largest in North America. The remote northern part of the island is the location of myriad caves, fissures and sinkholes. One particular cave, at the top of a steep forested valley, has proved to be a treasure trove of Ice-Age fossils. Both grizzly and black bear fossils have been found here. The grizzly proved to be a giant, as big as the Kodiak bears found today. He was dated at 35,000 years old, while the black bear was probably even older – over 40,000 years old. For these bears to have been here, the island, or at least a part of it, must have been free of ice during the Ice Age.

However the last Ice Age did not peak until 20,000 years ago, so perhaps glaciers covered the whole area at this later date. The fossil record suggests not. Further finds, including the jaw and ulna of a ringed seal, have been dated at 17,500 years old. Ringed seals could not have reached a cave at the top of a steep valley of their own accord – they must have been carried there overland. There is no way of doing that if the land is buried under 1.6 km (1 mile) of ice. One of the major

predators of ringed seals is the polar bear, but there is no evidence of polar bears ever having lived on Prince of Wales Island. A much smaller predator and scavenger of ringed seals is the arctic fox and, sure enough, the cave contains many small fossils of arctic foxes. Today, arctic foxes live near coastal areas on the open tundra of the north. Their presence on Prince of Wales Island at a time when huge ice sheets covered large parts of continental North America tells us that the coasts may have been an important ice-free refuge for some animals.

Recent research on grizzly bears living on Admiralty Island, just north of Prince of Wales Island, supports the case for the northwest coast having large pockets of ice-free land during the peak of the last Ice Age. The island is home to the Admiralty Island bear – a coastal brown bear or grizzly. Summer brings the bears down to the tidal mudflats where migrating salmon begin their long, arduous journey upstream. As tides ebb and flow, many salmon get trapped in the remaining pockets of water, giving the bears easy, and nourishing, pickings. Like most brown bears, Admiralty Island bears show huge variation in the shade of their fur. Yet even to the untrained eye, Admiralty bears look different from their mainland counterparts: many are a much deeper brown, almost black.

Analysis of the Admiralty Island bears has demonstrated that there is a clear genetic difference from mainland brown bears. These sorts of differences often

Left: The Arctic fox, today a major predator and scavenger in the far north of the continent, lived on Prince of Wales Island during the last Ice Age. **Right:** During the summer months coastal brown bears boost their protein intake by fishing on the tidal flats and digging for shellfish, particularly clams.

appear in species of animals isolated on islands for long periods. The level of genetic difference seen in the Admiralty Island brown bears suggests that these bears have lived in isolation for tens of thousands of years. If ice had covered the whole of Admiralty Island during the last Ice Age then this long period of isolation would have been interrupted. New bears would have had to colonize the island after the ice retreated. The genetic evidence tells a different story – Admiralty Island, or at least a substantial part of it, must have been free of ice and was a home to bears and other animals living in the shadow of huge glaciers.

WHEN IS A GRIZZLY NOT A GRIZZLY?

The term 'brown bear' is often used interchangeably with grizzly. All brown bears of North America come under the banner of a single species, *Ursus arctos*. However, some people believe that true grizzlies are found in the interior, while those on the coast are simply coastal brown bears. Grizzlies and Alaskan brown bears have certainly been separated out into different subspecies.

The word grizzly sounds ferocious but in fact may be a reference to the grizzled appearance of the bear's fur, which is lighter in colour at the tips. The brown bears that live in coastal regions tend to be larger than interior brown bears (or grizzlies). This is because they have a rich protein diet thanks to the huge migrations of fish, such as salmon, along the coast. Grizzlies have more of a plant-based diet and are consequently smaller.

Most brown bears hibernate, although in some coastal regions many bears make use of the milder climate to hunt for food all year round. The biggest of the brown bears is the Kodiak, which weighs in at a hefty 680 kg (1500 pounds). The largest modern bear, the polar bear, is believed to have evolved from a brown bear ancestor less than 1 million years ago.

Left: The first people in the northwest may have travelled between islands using boats. **Right:** The caves on Prince of Wales Island were used by people at the very end of the last Ice Age.

THE ANCIENT MARINER

The arguments surrounding the full extent of the ice sheets may seem a pedantic geological detail – so what if southeast Alaska was ice-free during the last Ice Age? However, these discoveries do have far-reaching consequences. They support the growing belief that the coasts, and not the interior, were the first places in North America to be occupied by people. If bears were living in this area during the peak of the last Ice Age then surely people could have been here too. Humans – like bears – are omnivores, able to live off a varied diet of fish, meat and plants. On Prince of Wales Island, further digs in the cave where the bear fossils were discovered led to a long-awaited find – the remains of a young man, believed to be in his early twenties. His lower jaw, part of his hip and some backbones were recovered. They proved to be post-Ice Age at around 10,000 years old. So, whether or not people actually made it to this island during the last Ice Age, they were clearly present very soon afterwards.

Analysis of the bones has told us a lot about this young man. The positioning of his wisdom teeth was typical of a man in his early twenties, but his molars were

heavily pitted for someone so young. This pitting is typically found in the teeth of people who consume large quantities of protein-rich (and gritty) shellfish in their diet. Chemical analysis of bones can reveal much about a person's diet and in this case it supports the argument that the young man lived on a marine diet, probably eating fish, shellfish and sea mammals.

The man on Prince of Wales Island was probably descended from people who arrived in the area much earlier. The degree of human-related material found around the cave hints that there may have been a camp there used regularly by people long before the young man died. Charcoal from fires, flakes from stone tool-making, and even ochre, a reddish pigment perhaps used on faces or bodies, have all been collected in archaeological digs over several years.

How did the man on Prince of Wales Island die? This is still uncertain but tiny scratches found on his pelvis were almost certainly made by a large carnivore, possibly a bear. Given the discovery of fossilized remains of bears within the cave, it may be that the young man had a struggle with a bear, perhaps hunting it during its hibernation. On the other hand he may have died from another unknown cause and simply have been scavenged by a hungry bear.

Thousands of forested islands lie off the coast of southeast Alaska.

The worked stone tools around the cave campsite were hand-napped from obsidian, a type of volcanic rock. This stone does not exist on Prince of Wales Island but can be found on the mainland around northern British Columbia. So people must have transported the stone to Prince of Wales Island, or it is possible that they acquired the rock by trading with other groups on neighbouring islands. Today the northern tip of Prince of Wales Island is surrounded on three sides by water. There is nothing to suggest that it would have been different 10,000 years ago when the young man hunted and fished the coasts. Somehow he and others like him were crossing the water between the islands.

SEA TRAVEL

The quickest and most efficient way to do this would have been to use some kind of boat or watercraft. There is, as yet, no direct evidence of boats being used in North America as long ago as the end of the last Ice Age. Although the evidence has still to be found, the use of boats to colonize a continent is not without precedent. On the other side of the Pacific, the continent of Australia has been isolated from Asia for millions of years. However, archaeological evidence points to people arriving on this continent around 50,000 years ago. These people must have used a watercraft of some sort, perhaps hopping via islands from the Asian mainland to Australia itself. If people were able to cross water 50,000 years ago, the use of boats as a means of entry into North America 14,000 years ago, or possibly earlier, seems highly plausible.

NATURE'S SEAFOOD PLATTER

Undoubtedly the marine environment would have offered a varied, and relatively accessible, diet to early people travelling into North America. They may have followed and hunted sea lions. Steller's sea lion, for example, covers hundreds of kilometres in search of food. Sea lions would also have drawn people to the northwest's biggest fish resources – herring spawning and salmon migration.

Salmon spend most of their lives at sea before returning to the rivers where they were conceived. It is a one-way journey, for in the process of spawning and starting the next generation they die. A variety of salmon migrate to the rivers of the northwest, including silver, king and sockeye. King salmon is the largest, weighing up to 59 kg (130 pounds) and spawning at around four or five years of age. Spawning events are spread throughout the year, though early summer and autumn see peak numbers entering the rivers. This attracts a variety of

KAYAKS, CANOES AND UMIAKS

The coast-inhabiting indigenous people of North America have a long cultural history of boat making that can be traced back thousands of years. Different types of watercraft have evolved, depending on locality and family tradition.

Inuits use the term 'kayak' for the fast, slender skin-covered boats that are used for hunting marine mammals, fishing and transporting ice. Kayaks are long and narrow with a hatch or manhole in which the kayaker sits. Harpoons, clubs and hooks would have been transported on the skin deck. Sealskin was traditionally used in the construction of these boats, but nowadays synthetic materials have largely taken its place.

Canoes are more open and are often used to carry several people. Traditional canoes were made using bark or cut in one piece from a tree trunk. The northwest is particularly rich in the durable wood of red cedars. Canoes were mainly used for transport along rivers and for fishing.

The umiak, like the kayak, is traditionally a skin-covered boat. It is an open boat with a wide base for carrying large numbers of people and equipment. In the past, whole families may have made seasonal journeys in a umiak. They were also used for whale hunting.

Salmon travelling upstream to spawn may well have drawn the first people inland from the coasts.

predators, but one of the greatest beneficiaries of the salmon migration has to be the brown bear. During the last Ice Age, salmon would have found their normal spawning rivers covered by ice. Yet it seems that the Columbia River, one of the biggest rivers in this area, remained free of ice and would have witnessed huge numbers of salmon migrants. It is also thought that some rivers continued to flow under the ice and that salmon may still have utilized them, perhaps even moving up through cave rivers.

Pacific herring school in large numbers off much of the northwest coast. Their silver appearance produces a dazzling show when they gather in their thousands and move in synchronous response to the threat of a predator. They come into bays and estuaries to spawn; in the northwest this is usually in early spring, between February and April. Herring are much smaller than salmon, at around 46 cm (18 inches) long, but they tend to live longer – up to 11 years. Salmon are a major predator of herring, as are seals and sea lions. The herring spawning event is spectacular, with females releasing over 100,000 eggs, which usually attach to marine plants and rocks. The males are equally profligate with their sperm, turning huge areas of water milky in the process.

Spawning events may have provided bumper catches for the first North Americans but they are highly seasonal. More regular pickings would have been made in the intertidal zone, which becomes exposed during low tides. There is a saying that on a beach the table is set twice a day. The earliest North Americans are often depicted as spear-wielding mammoth hunters. While there is no doubt that they hunted big mammals, protein in the form of shellfish would have been much more readily accessible in coastal areas during low tides. Men and women could have carried out the task of collecting shellfish, but it may well have been a chance for the older and younger members of a group to supply food for the dinner table. There is even evidence that people wove baskets using plant fibres, perhaps to carry the food they had collected on foraging trips along the coasts and rivers.

Resourcefulness and adaptability have long been the keys to human survival. Living on the coasts required the combined skills of fishing, hunting and foraging. At some point, people entered the interior of the continent, eventually reaching the open grasslands of the great plains. Whether this was through an ice-free corridor or via the coasts, they would have encountered a very different landscape, richly populated with animals never before witnessed by people.

The northwest at the end of the last Ice Age: an old bull mastodon lies on its side, close to death, but is protected by another mastodon, ready to see off the potential threat of a nearby scimitar-toothed cat.

THE GREAT PLAINS

Left: The great plains of North America were once home to vast herds of bison. **Above right:** The rolling grasslands and wooded gullies of the northern plains in Alberta, Canada. **Below centre:** A bull bison can weigh over 1000 kg (0.9 tons), making it the largest land mammal in North America today. **Below right:** Prairie dogs are the keystone species of the prairie grasslands.

AMERICA'S SERENGETI

The prairie grasslands of North America cover an immense area of land, stretching some 3800 km (2360 miles) from Mexico into central Canada and from the foothills of the Rocky Mountains east for nearly 1600 km (995 miles) to the Mississippi valley and beyond.

As recently as 200 years ago, the prairies were one of the richest grasslands ever to have existed on Earth. This region supported an almost unimaginable abundance and diversity of animals. Single herds of bison and pronghorn may have numbered several million animals, stretching from horizon to horizon as they drifted across the plains. Millions of prairie dogs and gophers burrowed through the rich soil, improving the grazing for these vast herds and repairing the land after they had passed. Packs of wolves hunted the bison, picking off the vulnerable young or injured, while marauding grizzly bears could take their pick from the feast. Each spring and autumn the sky would darken as clouds of migrating ducks, geese and cranes made their seasonal migrations.

Following the arrival of European settlers during the nineteenth century, it took only a few decades for this seemingly endless abundance to be reduced to a pitiful shadow of its former greatness. The estimated 65 million bison were hunted to virtual extinction and the pronghorn antelope almost vanished, its great migration patterns cut by fences and settlements. The last wolves and grizzly bears disappeared from the prairies in the early twentieth century.

These are just the last in a long line of dramatic changes the plains have gone through. It was not the first time that these grasslands have played host to an amazing variety of animal life, nor the first time that new arrivals have radically altered the appearance of the landscape. Around 13,000 years ago, the first people arrived in this region. They found it teeming with life: a true American Serengeti. So how did the plains appear to these pioneers and what kinds of animals did they encounter?

The aspen parklands region in southern Alberta, Canada, resembles the way much of the northern plains may have looked towards the end of the last Ice Age.

FALLING INTO THE PAST

We can get an idea of the richness and diversity of the plains during the last Ice Age from a site tucked away on the western edge of the Bighorn Mountains in northern Wyoming. Natural Trap Cave is a gaping hole in the ground with a narrow shaft that drops some 25 m (82 feet) into the cavern below. This cave has been 'collecting' animals for over 20,000 years. Locked away in the sediments of the cave floor are the remains of many creatures unfortunate enough to fall into the trap. Some of these are familiar animals, still found in the area today: antelopes, rabbits, gophers and various rodents. Older sediments, dating from 10,000 years ago and earlier, contain creatures long departed from this desolate landscape – animals such as dire wolves, short-faced bears, lions, cheetahs, mammoths, horses, camels, woodland muskoxen, as well as extinct species of bighorn sheep, bison and pine marten.

The circumstances that led these animals to fall into the cave remain obscure. Perhaps they were being chased by a predator or were in headlong pursuit of prey. Maybe it was dark or snow was falling, and they accidentally tumbled over the edge. Whatever the reason, they all suffered the same fate. If the fall did not kill them outright, they would have found themselves trapped far below ground with no escape route. Some of the more poignant finds in the cave include scratch marks on the cavern walls, which show where trapped and injured animals made desperate and ultimately futile attempts to climb out of their underground prison. The discovery of animals such as collared lemmings, caribou, muskoxen and arctic hares in the most ancient sediments suggests that a far colder climate once prevailed in the plains. Today these animals are only found in much higher latitudes (see Chapter 1, p. 13). It is only in the sediments dating from 15,000 years onwards that temperate-climate creatures come to prominence.

There is further telling evidence for the colder, distant past of the great plains if you know where to look. Clues can be found in the shape of the land, in the drainage patterns of the great rivers, in the myriad ponds and lakes that dot the plains and even in the make-up of the soils.

AFTER THE ICE

The great plains started life as an ocean floor. For the best part of 500 million years, seas advanced and retreated across it, laying down seabeds of limestone, shale and sandstone. About 65 million years ago, at the end of the Cretaceous period, this ocean floor was exposed and tectonic action began to create the Rocky Mountains along the western edge of the plain. Since then, millions of years of erosion have washed sediments out of these mountains and onto the flat lands to the east. The result is that the once flat seabed is now a gentle 1300 km (800 mile) long debris slope, which runs down to the Mississippi Valley.

Subsequently, the plains have been shaped and reshaped by repeated coverings of ice, the last of which advanced southwards from the Arctic around 28,000 years ago. At the peak of the last Ice Age, some 20,000 years ago, the ice sheets were over 1.6 km (1 mile) deep. They stretched south from the north pole to a line approximating the present US–Canadian border, with extensions spreading south along higher ground like the Rockies. Although the ice sheets only lapped against the northern parts of the great plains, their effects were felt across the

The prairie pothole region in North Dakota. Today the region is known as the 'duck factory' as over 75 per cent of all North American wildfowl originates here.

entire region. They modified the surface of the lands they covered, and deposited enormous quantities of debris over the plains as a whole. Thick layers of sediments, from fine silt to sand and even huge boulders, were picked up by the moving ice and deposited far from their place of origin. These deposits – glacial soils and rocky moraines – are the foundation for many of the current landscape features on the plains. Only with the rise in global temperatures did the ice sheets finally retreat north.

PRAIRIE POTHOLES

As the ice sheets retreated northwards, isolated blocks of ice, some huge, were left behind in the sea of rocky debris. Insulated by a covering of dirt and rocks, these ice blocks persisted long after the main body of ice had retreated. However, slowly the heat of the sun penetrated them and as they melted the 'ground' above collapsed, leaving a series of craters that subsequently filled with water. This created a landscape dotted with thousands of lakes of all sizes – the so-called prairie potholes – which are mainly found in the Missouri and Prairie Coteau regions of the Dakotas.

A prairie dog grazes on the vegetation around its raised burrow entrance. Some is eaten, some simply clipped to improve the chances of spotting approaching predators.

The sediments that lie at the bottom of prairie ponds act as a time capsule, holding all kinds of clues about how the plains once looked. Each spring, pollen from the surrounding plants falls onto the ponds, then sinks and is buried at the bottom. Over the years, alternating layers of pollen and silt build up, recording for ever the types of pollen falling onto the water. Like the annual rings found in a tree, this information acts as a window on the past. Using samples from ponds across the plains, it is possible to build up a detailed picture of the plants that once grew across this region.

They tell us that when the first people arrived here the ice sheets were already in retreat. The land south of the ice front was bleak. The freezing temperatures, incessant winds and disrupted drainage encouraged only tundra, muskeg bog and boreal forest. But as the climate continued to warm and the ice sheets carried on melting, vegetation zones previously confined to the warm south began to spread northwards. Encouraged by the cool, moist summers and mild winters, trees flourished across much of the plains, creating a vast parkland of conifers and aspens interspersed with prairie-like meadows.

As the climate warmed further, this parkland in turn began to disappear, to be replaced by an endless 'sea' of prairie grasses – a situation that prevailed until the European settlers began to farm the plains and plough up the rich grasslands during the nineteenth century.

The vast prairies were not completely uniform: bands of different grasses spread across the plains depending on the amount of summer rainfall. The lands nestling up against the Rockies receive as little as 25 cm (10 inches) of rain a year, and this created an almost arid strip that supported only scrubby grasses, cacti and other desert plants. Further east, the land receives more rain, encouraging a richer growth of grasses such as little bluestem, wheatgrass and prairie june-grass – the mixed-grass prairie. Further east still, yet more rain falls, nurturing tall grasses over 2 m (6.5 feet) high. Today, only small remnants of this tall-grass prairie survive.

PRAIRIE DOGS

The vast treeless plains create problems for the animals that inhabit them. Without trees, there is little shelter from the elements or from predators. Animals living here must either be able to run, like pronghorn antelopes, bison and jackrabbits, or be able to dig, like the black-tailed prairie dog.

Prairie dogs are not dogs at all, but highly social rodents. They live in large, densely packed colonies or 'towns', which often cover many square kilometres. The towns contain thousands of burrow entrances, up to 20 per hectare (50 per acre), which lead into a labyrinth of tunnels that can extend 5 m (16.4 feet) underground.

These tunnels give sanctuary from predators and from the fires that sweep across the plains each summer. Thirteen-lined ground squirrels, mice and voles, as well as burrowing owls, all nest in vacant burrows. Rattlesnakes hibernate in them, spiders weave webs across the entrances of deserted holes, and insects such as dung beetles, rove beetles and fly larvae live within them.

The prairie dog has several important functions in the local ecosystem. Their constant burrowing aerates, and their dung fertilizes, the soils. Their continual pruning of the vegetation around the burrows encourages the growth of both grasses and other plants. This in turn provides highly nutritious forage that is irresistible to the larger grazing animals like bison and pronghorn antelopes. During the spring, it is common to see herds of both species occupying prairie dog towns.

Finally, prairie dogs are prey for virtually all the other prairie hunters. Coyotes, badgers, black-footed ferrets, rattlesnakes, golden eagles and many other raptors all hunt the dogs.

THE GRASS IS ALWAYS GREENER

For thousands of years the wide-open plains were dominated by one animal: the bison. Until they were decimated by hunting in the last 200 years, these huge creatures – the largest of all North American land mammals alive today – roamed the grassy plains in herds that are thought to have numbered tens of millions (see Chapter 6, p 180).

However, the mosaic of grassland, parkland and forest that covered the plains 13,000 years ago would have made such huge gatherings of animals rare at that time. Instead, the fragmentation of the habitat is likely to have favoured smaller groups of bison. These groups may have consisted of just tens of animals: mothers with their young, and a few attendant males fighting for females during the breeding season. The animals would have migrated seasonally, seeking fresh food supplies by following the rains. They may well have come together in larger groups in the summer, during the breeding season, when grasses and forage would have been more plentiful.

But 13,000 years ago bison were not the largest creatures found here – far from it. For evidence of this we must once again look underground.

A MAMMOTH GRAVEYARD

Some fossil finds are the result of intense scientific efforts; others come to light by sheer chance. In 1974, while the foundations of a new housing estate were being dug in Hot Springs, South Dakota, the earth-moving equipment uncovered some unusual fossilized bone fragments, including what appeared to be an elephant's tusk. In fact, after closer analysis, it was discovered that these bones belonged to a Columbian mammoth, the largest of all the ancient elephants that once roamed North America.

The building workers had stumbled upon one of the most important fossil sites in North America. So far, over 50 fossilized mammoth skeletons have slowly and painstakingly been uncovered at the Hot Springs site, which covers an area of 50 m (164 feet) in diameter. It is thought that another 50 skeletons may still lie buried there.

A herd of bison grazes the summer grasses. The young are born in the spring and can run within hours of birth, enabling them to keep up with the herds on their constant search for fresh grazing.

Left: Hot Springs, South Dakota. The skeletons of over 50 mammoths have been excavated from the sediments of this Ice-Age pool.
Right: The study of African elephants has revealed much about how Ice-Age mammoths must have lived.

How did so many mammoths come to perish in such a small area? The excavation site is an ancient, spring-fed pool, which existed for several hundred years in the late Pleistocene era, around 20,000 years ago. This spring would have been attractive to mammoths, especially in the cold winter months, because of its warmth and feeding opportunities. Mammoths venturing into the pool or foraging around its edges may have found they could not climb back up the steep, slippery banks. Trapped, they either drowned in the deep waters or slowly starved to death. Over the centuries, many mammoths met their end in this watery grave until finally the pool filled with sediments, burying the bones.

We can determine the sex and age of the mammoths from their bones and tusks. Remarkably, all the mammoths found so far at the Hot Springs site have turned out to be young males. An intriguing explanation for this has come from studies of elephants in Africa. Here, there is a higher mortality rate among young males than any other group. This is because they are forced to leave the herd at around 12–15 years of age, and these single roving bulls are more adventurous than other elephants and the most likely to find themselves in dangerous situations. Females on the other hand live in larger groups, are more cautious and can help each other out of trouble. It was perhaps the same in Columbian mammoth society.

SLAUGHTER OF THE INNOCENTS

Further insights into the social life of Columbian mammoths can be gleaned from the Dent site in Colorado, where in 1932 the bones of several mammoths were found along with some spear points. This suggested that the herd had been slaughtered by a group of human hunters and that most of the mammoths, if not all, had been killed at the same time. Studies of the tusks and cheek teeth subsequently confirmed that all the animals died in the autumn, supporting the idea that they died together. This is particularly significant because if the whole herd was killed at one time, then its composition can be determined from the remains.

In all, 13 animals were identified – eight juveniles and five adults. Based on age-estimating techniques, four of the adults were reckoned to be between 22 and 28 years and one 43 years old. Of the juveniles, there was one 2-year-old, two 3-year-olds, one 6-year-old, one 9-year-old, two 10-year-olds and one 14-year-old. Based on skeletal measurements, the adults were all thought to be females and the juveniles a mix of males and females. This herd composition, with a large female matriarch, several mature females and their sexually immature offspring of both sexes, is exactly mirrored in contemporary elephant society.

DWARF MAMMOTHS

Living on islands can do strange things to animals. Faced with very different circumstances from those on the mainland, animal populations isolated on islands can quickly change. One of the most obvious and well-documented changes is in body size. Many Pleistocene mammals isolated on islands became dwarfed.

Between 30,000 and 12,000 years ago, the flyweights of the elephant world lived on the Channel Islands in the Gulf of California. Called *Mammouthus exilis*, they were dwarf versions of the mighty Columbian mammoth. Standing at only 1.2–1.8 m (4–6 feet) tall at the shoulder, these dwarf mammoths were half the height of their mainland cousins and weighed about 1000 kg (0.9 tons), compared to the 10,000 kg (9.8 tons) of the average Columbian mammoth. The remains of over 50 of these dwarves have so far been recovered, mostly on the island of Santa Rosa.

Dwarfing is thought to occur in places where food resources are restricted and there is an absence of predators. In a limited amount of terrain, food is at a premium and small-bodied animals that can make do with less food should survive better, especially in times of shortage.

How did these mammoths come to be on the islands in the first place? During the last Ice Age sea levels were much lower than today – not low enough to link the islands to the mainland, but bringing them close enough for mammoths to have swum there. We should not be too surprised as elephants today often swim long distances.

CLASH OF THE TITANS

The Hot Springs site has informed us about young bachelor mammoths and the Dent site about the structure of matriarch-led female herds. But what about adult males? A remarkable find made in the 1960s near the town of Crawford in northwest Nebraska has filled this gap in our understanding perfectly.

Nicknamed the 'Clash of the Titans', the find consisted of two massive bull-mammoth skulls with their tusks locked together. Each mammoth had one full tusk and one broken tusk. This asymmetry may explain how the two came to be locked together. When elephants with two complete tusks fight they are kept at a distance by the curve of their tusks and the bulk of their trunk. Adversaries with broken tusks have no such buffer and can get much closer to each other, using the good tusk like a spear for stabbing.

Around 12,000 years ago this is exactly what must have happened when these two mammoths squared off against each other. As they pushed and shoved,

Left: Unearthed in Crawford, Nebraska, in 1962, these interlocked skulls are evidence that male mammoths must have fought over access to females in much the same way as African elephant bulls do today. **Right:** During the fight the tusks became locked together and the animals died face-to-face.

one mammoth punched a hole through the shoulder blade of his opponent, probably using his broken tusk. As the fight continued, the tusks somehow became irreversibly interlocked. The animals no doubt struggled in this state to the point of exhaustion, when they collapsed, dragging each other down. Locked in a fatal embrace, they eventually died.

Studies of the bones and teeth suggest that these males were about 40 years old when they died, an age when modern bull elephants reach full sexual maturity. This may explain why the mammoths were fighting in the first place. Forced out of the family herd at around 12–15 years of age, male African elephants wander the plains until they reach sexual maturity. They then only associate with female herds during the breeding period, at which time they enter a condition known as 'musth' when they become very aggressive and will battle each other for the chance to mate with any receptive female. Most fights are settled by intimidation and pushing contests but occasionally, when two males are evenly matched, the confrontation can escalate into out-and-out fighting. The evidence suggests the Crawford mammoths were full adults in the prime of life and well matched.

COLUMBIAN MAMMOTH

KEY FACTS

Common name: Columbian mammoth

Scientific name: *Mammuthus columbi*

Size: Over 4 m (13 feet) tall at the shoulder

Weight: 10,000 kg (9.8 tons)

Diet: Vegetarian – grazing on grasses and also eating a wide variety of different plants and browsing on conifers

Habitat: Savannah, grasslands, parklands and deserts

In a land of giants, one creature stood head and shoulders above the rest – the Columbian mammoth. The greatest of all herbivores on the plains, the Columbian mammoth was one of the largest elephants ever to have lived, bigger by some margin than the African elephants of today. A full-grown male was a huge beast, standing over 4 m (13 feet) tall at the shoulder and weighing a massive 10,000 kg (9.8 tons). The Columbian mammoths were close relatives of the woolly mammoth (see Chapter 1, p. 20) and were found mainly in areas covered in savannah, grasslands or tundra.

Columbian mammoths had an appetite to match their size. Based on studies of elephants, it is estimated that an animal this size would need to consume around 250 kg (550 pounds) of grassy vegetation each day and spend up to 20 hours a day foraging for food. This in turn meant the mammoths had to cover great distances in order to find sufficient forage. They may have followed migration paths that tracked the growth of fresh grasses, much as wildebeest and other animals do in the African Serengeti today.

Many of the fossilized jaws of Columbian mammoths still contain teeth, which reveal important information about the diets of these huge animals. Just like modern elephants, mammoths had four molars, both upper and lower, on each side of the jaw. The surfaces of these huge teeth were very rough, not unlike corrugated cardboard. Studies of pollen and other debris trapped in these ridges, and between the teeth, show that the Columbian mammoths ate mainly grasses and sedges.

These huge molars acted as grinders, shredding the coarse vegetation. The teeth had to cope with a huge amount of food each day and were incredibly resistant to wear and tear. Nevertheless, grass is hard to digest: it contains tiny particles of silicon, which can act like sandpaper. These particles slowly wore down the teeth until the ridges were smoothed off. But when the teeth were worn flat they simply got pushed out by new teeth growing behind. In total, a mammoth had six sets of teeth to get through in its lifetime, the last set usually coming through at around 30 years of age.

Grass is a tough food and even with such massive grinding teeth only about half of it was actually digested. The remainder passed straight through. Of the 250 kg (550 pounds) processed each day, some 90 kg (200 pounds) of dung was produced.

Molars were not the only specialized teeth found in mammoths. Just like modern African elephants, Columbian mammoths also had tusks. And they were massive. In the American Museum of Natural History in New York there is a set measuring 3.5 m (11 feet), while other finds have exceeded 4.5 m (15 feet). The tusks would have continued growing throughout the life of a mammoth,

Columbian mammoths were one of the largest elephant species ever to have lived. An adult bull weighed over 10,000 kg (9.8 tons) and needed to eat 250 kg (550 pounds) of food every day. Based on studies of African elephants, this would have meant feeding for about 20 hours each day.

Ice-Age distribution of Columbian mammoth

starting as stubs just a few centimetres long in new-born calves and growing at a rate of anything up to 15 cm (6 inches) a year. Unlike with modern elephants, the tusks of a mammoth did not grow in a simple curve but formed a spiral or corkscrew shape with the tips often overlapping. They would have been used in both fighting and feeding.

It is possible to make certain generalizations about the social life of Columbian mammoths based on the many fossils finds unearthed across North America. In addition, comparisons with modern elephants are a key source of information. Columbian mammoths probably lived for around 60 years or more. Social life revolved around the female herds, with groups of related individuals staying together throughout their lives. Herds would have numbered between 2 and 20 individuals, led by a dominant 'matriarch', and comprising a number of other adult females and their offspring. Gestation was 22 months, after which a single young was produced, which would have been suckled until it was two or three years old. Female young would stay with the herd while males would leave when they reached 12–15 years of age. Adult males lived apart from the herds, joining them only during the breeding season to mate with receptive females. Adult males would have fought for access to females at this time.

THE ROCKET OF THE WEST

Many species that existed during the last Ice Age are still present in the region. One of the most recognizable animals on the plains today is the pronghorn antelope, a creature unique to North America. Incredibly tough, it can go for many days without water and will eat and thrive on plants no other grazing animals will touch – even cacti.

The pronghorn's main claim to fame is that it is the fastest animal in North America. It possesses an impressive ability to maintain phenomenal speeds over considerable distances. Officially clocked at 96 km/h (60 mph) for several minutes, pronghorns can run at speeds in excess of 112 km/h (70 mph) for short bursts. Young pronghorns just a few days old are able to run as fast as 72 km/h (45 mph).

The key to maintaining such speeds is the ability to get oxygen to the muscles quickly. Pronghorns are uniquely adapted to this end. With an oversized wind-pipe that sucks in huge quantities of air as they run, they consume about three times the amount of oxygen of similar-sized animals. Pronghorns run with their mouths open, forcing air into their huge lungs, and they have a massive heart to pump this super-rich blood around their body. To cope with the speed, the pronghorn's leg bones are extremely strong and their hooves are padded to minimize shock.

Why do pronghorns run so fast? Today, the fastest predator they ever have to confront is the wolf. However, while wolves working in relay can pursue prey for many kilometres, they do not run particularly fast. Clearly pronghorns can both out-sprint and out-distance them. So why do they have such athletic abilities?

Fossil finds from across the plains help to explain this conundrum. During the last Ice Age, pronghorns had to contend with far speedier adversaries – cheetahs. Larger, leggier and probably faster than their modern-day African cousins, American cheetahs must have been fearsome hunters. Extraordinary speed was the only defence against them. Perhaps the pronghorn's physiology was adapted to escape from this predator and today they continue to possess attributes that were necessary in their past.

Pronghorn antelopes are the fastest North American land mammal, capable of speeds in excess of 96 km/h (60 mph).

THE RIDDLE OF THE ROTTING FRUITS

There are several North American tree species that produce huge pulpy fruits only for them to fall to the ground and rot. Plants such as the honey locust, Osage orange, desert gourd and Kentucky coffee have large fruit crops that remain essentially untouched by any animal.

So why go to all that effort for nothing? A major clue comes from the fruits themselves. The fruit of the Osage orange is the size of a melon – much too large to be eaten by any animal alive in North America today.

However, 13,000 years ago there were several creatures that could easily have consumed such large fruits – animals such as mammoths, mastodons and giant ground sloths. When these giants disappeared, there were no animals to disperse Osage seeds and so the plant went into decline.

Today, natural stands of Osage are few and far between. It is only the 'unnatural' intervention of farmers, who have planted the thorny plant into hedgerows, that has saved the Osage.

Cheetahs once hunted across the Ice-Age plains. Larger than their African cousins, they would have preyed on the smaller plains animals such as pronghorn and deer.

PRAIRIE HUNTERS

The Ice-Age grasslands were home to an abundance of hunters. American cheetahs rubbed shoulders with more familiar North American predators, such as coyotes and mountain lions, grizzly bears and wolves. Together they formed perhaps the most fearsome collection of hunters ever gathered in one place.

Buried deep inside the hills of southern Missouri, Cat Track Cave contains a remarkable piece of evidence – not the usual fragments of bone or even a skull, but a far more intimate link to the past. In the soft mud of the cave floor, a passing animal has left a set of footprints. Clearly cat-like in shape, what is immediately striking about the tracks is their sheer size – about 17 cm (7 inches) across. This was no wolf or a mountain lion. The prints are as large as those made by the Siberian tiger, the largest cat alive today. They were in fact left by an American lion, a creature that vanished from North America at least 12,000 years ago (see Chapter 1, p. 33). These big cats, which probably hunted in small prides, must have been an intimidating sight. They could have easily tackled huge and awkward prey such as bison and camels.

The lions shared the Ice-Age plains with, among others, jaguars. These solitary cats were found throughout the region at that time and their remains have been recovered from caves in Missouri and elsewhere.

Dire wolves too were widespread. Although very similar to grey wolves, they were substantially larger, with a more powerful appearance. They had a large head, massive teeth and short, muscular legs. The greater number of dire wolf skeletons relative to other hunters found in the La Brea tar pits in California (see p. 95) suggests that they were highly social creatures, just like modern wolves. They would probably have worked together to hunt and scavenge, and their powerful jaw muscles would have enabled them to crack open bones much as hyenas do today. Their teeth would also have been good for hanging onto larger prey, such as horses, camels, deer and bison. Several wolf skulls from La Brea show signs of injury from powerful kicks, probably sustained as they chased after their victims.

This is already a formidable cast of hunters. But there was one other group of predators stalking the plains 13,000 years ago, perhaps the most menacing of all carnivores – the sabre- and scimitar-toothed cats.

SABRE-TOOTHED CAT

KEY FACTS

Common name: Sabre-toothed cat

Scientific name: *Smilodon fatalis*

Size: 1 m (3 feet) at the shoulder, 2 m (6.5 feet) from rump to snout

Weight: 300 kg (660 pounds)

Diet: Carnivorous – horses, camels, bison, antelopes and deer. May have also eaten carrion.

Habitat: Grassy plains, open woodlands, deserts

Found in both North and South America, the sabre-toothed cat is arguably the most charismatic of all the Ice-Age creatures. The outstanding finds in the La Brea tar pits in California (see p. 95) have provided an unrivalled opportunity to study *Smilodon fatalis* and have made it one of the best known of all the fossil cats.

Structurally, the sabre-tooth's skeleton was similar to that of a modern lion. But at around 1 m (3 feet) tall at the shoulder and 2 m (6.5 feet) from snout to rump, this was a much smaller animal. However, at 300 kg (660 pounds) it was close to double the weight of a lion, making it a very squat, powerful beast. It also had a bobtail, unlike lions. The sabre-tooth's most instantly recognizable feature was of course its long, laterally compressed upper canines – its sabres. At 24 cm (9 inches) long, these were among the most impressive killing weapons in the animal kingdom.

Several functions have been ascribed to the sabres. It has been argued that they were used for climbing trees, grubbing for food through sand and soil, or piercing the skull of a victim in the way jaguars do today. So how would a sabre-toothed cat really have used these rather unwieldy weapons?

The sabres themselves were relatively blunt. While capable of penetrating skin and muscle given sufficient force, there is a good chance they would have broken if they had hit a bone. Simply sinking the teeth deep into their victim's flesh would have been a risk to the sabre-tooth as any struggling by the prey would have threatened to break the teeth.

It is possible that the sabres were driven into the prey and withdrawn quickly, ripping flesh and blood vessels in the process. The most obvious point of contact for such a bite would be the neck, where it could do maximum damage to the windpipe and major blood vessels. The cat could then retire while its victim died from blood loss and shock. This type of bite would have to be delivered to prey that was held down and immobile, and this is where the cat's great bulk came into play.

But how did the sabre-tooth catch prey in the first place? No tail, short legs and an extremely bulky body are hardly the features of a thoroughbred. The sabre-tooth could not run far or fast like a lion or cheetah. However, its weight meant that if did hit an animal at speed, the chances are its victim would go down. The hapless prey would then be pinned under the cat's great bulk as the death bite was administered.

This suggests that the sabre-tooth hunted large, lumbering prey – animals such as juvenile mastodons, camels or even young mammoths. Since it was not able to chase efficiently, the sabre-tooth was better off lying in wait for vulnerable prey to draw near and then ambushing its victim.

The sabre-tooth's large front teeth presented it with other problems. Once its prey

Centre: Sabre-toothed cats were powerfully built. Their neck and shoulders were especially strong to enable them to pin struggling animals to the ground as they drove their sabres into them. **Below right:** The sabre-tooth had an extraordinarily wide gape to allow the sabres to clear its bottom jaw – over 90 degrees compared with the 60 degrees of other cats.

Ice-Age distribution of sabre-toothed cat

was dead, how did the cat actually manage to eat anything? It would have to extend its mouth through a full 95 degrees to allow the sabres to clear its lower jaw. Most modern-day cats have a gape of around 65 degrees. The sabre-tooths were capable of this astonishing feat, but they actually overcame the difficulties presented by their front teeth by using specialized carnassial, or slicing, teeth at the back of the mouth. Meat was grasped and bitten off through the sides of the mouth while the cat was sideways on to the kill.

Sabre-toothed cats in La Brea were found to be three times more likely to have tooth fractures than equivalent carnivores today. When food is limited, carnivores are likely to feed more rapidly, guard their kills more aggressively and completely consume their prey, often ingesting bone in the process. All these activities increase the risk of tooth damage or breakage.

Tooth breakage rates suggest that Ice-Age hunters and scavengers would eat more of a carcass than their modern-day counterparts. This might have been because times were tough on the plains and prey was difficult to catch and retain. Or perhaps there was an absence of prey, at least seasonally, due to patterns of migration. Under these circumstances, carcasses would have become extremely valuable resources.

Left: The sabre-toothed cat was the ultimate Ice-Age predator, but it was no match for this skunk's evil-smelling defences.
Right: Sabre-tooths may have lived in social groups, as African lions do today.

SIZE MATTERS: THE SABRE-TOOTHS

Although sabre-toothed cats were found throughout much of the North American continent, most of what we know about them comes from one site – the La Brea tar pits in California. Over a 25,000-year period at least 2500 sabre-toothed cats were trapped here, and from the 166,000 bones found we can begin to picture the life of this famous predator.

One of the most contentious arguments about the sabre-toothed cat centres on whether it was a social or solitary animal. Evidence from an unexpected source at La Brea has shed some light on this debate. It is not the hundreds of perfect specimens, but the many mangled, broken and diseased bones pulled from the tar that have proved most telling. At least 5000 bones from La Brea show some kind of pathological condition. Some cats had dislocated hips, others had crushed vertebrae and many had leg bones hideously deformed by lumpy projections, which suggest that the cat's body had been trying to shore up areas under extreme pressure. Such traumatic injuries are most likely to have occurred when the cats were hunting or wrestling with heavy prey.

What is most revealing about these finds is that many of the damaged bones show signs of healing, even when the wounds were so severe that the cat would have been left permanently disabled. The evidence suggests that these animals managed to live for months or even years after sustaining their injuries. This is the crux of the debate. The injured cats almost certainly could not have hunted for themselves – so how did they survive?

One possible answer is that the sabre-toothed cats lived in social groups, like lions. Wounded animals were supported by other members of the group until they were healthy again. If not actively helped to recover, the injured may have been tolerated around kills and have been allowed to eat the scraps left by other cats.

However, the idea of social sabre-toothed cats is not uniformly accepted. There are other possible explanations for the survival of injured and maimed animals. La Brea was a place associated with dead animals, and the presence of injured sabre-toothed cats may simply mean they came here to scavenge from animals already trapped in the tar. Furthermore, even badly wounded cats would have been able to use their physical presence to intimidate other animals away from kills – especially smaller predators such as cheetahs and wolves.

LA BREA TAR PITS - A STICKY END

One of the most extraordinary fossil sites in the world is found in one of the least likely settings – downtown Los Angeles. Surrounded by skyscrapers and highways, the La Brea tar pits are an amazing window on the past. Over 100,000 kg (98 tons) of fossils, 1.5 million bones and 2.5 million invertebrate remains have so far been removed from the pits. Most date from 40,000–10,000 years ago and range in size from insects and birds to Ice-Age giants like ground sloths and mammoths. Bones from 3400 individual large mammals, including six different predator species, have been recovered.

How did so many animals end up here? The asphalt at La Brea seeps up to the surface from deep underground, especially in the hot summer months. It forms shallow 'puddles' that would have been hidden by a covering of leaves and dust. Unwary animals straying into the area would have become trapped on these thin sheets of extremely sticky tar.

Like flies caught on flypaper, the animals died of exhaustion and dehydration. As the animals decayed, more scavengers would have been attracted and similarly caught in the tar themselves. Surprisingly, far more carnivores and scavengers have been recovered from the tar than herbivores. For every large plant-eater that died here six meat-eaters met a similar fate.

THE ICE-FREE CORRIDOR

Around 13,000 years ago, another type of hunter arrived on the plains – humans. Working in groups, armed with formidable weapons and using sophisticated hunting techniques, these highly efficient predators would quickly come to dominate the grasslands and indeed the whole continent. Whatever route the first people used to arrive south of the ice – either via the coasts and then over the Rockies, or through the newly opened ice-free corridor – it was clear that they were poised to change this American Serengeti for ever. With their weapons and generations of knowledge and experience gained while hunting in Asia, humans were destined to have a profound effect on the plains and its animals. Creatures that had never encountered them before must easily have fallen prey to their weapons and tactics.

The Colby site, near the town of Worland in the Bighorn Mountains of central Wyoming, represents one of the largest mammoth kill sites so far discovered in

Left: Bison were hunted by the first people and bison bones have been found in association with many human butchering sites across the plains. **Right:** The people inhabiting the Ice-Age plains were expert hunters. It is thought that excess food from a kill was stored in caches for later retrieval, and there is some evidence to suggest that these larders were marked with the bones of prey.

North America. The find includes two distinct piles of bones, which number over 450 in total. The majority of bones belong to at least seven immature Columbian mammoths and one foetus. Bones of rabbit, ass, pronghorn, camel and bison have also been found at the site.

The position and types of bones (especially the mammoth bones) suggest that the animals were butchered and preserved in two large caches, or meat stores. The bones were piled high and then covered with rocks to protect them. Several spear points were also found with the cached meat. Animal skulls appear to have been placed on the top of each cache, perhaps as a marker.

However they arrived here, these first people found themselves poised to exploit the lands to the south of the ice sheets. Following the great rivers that drain the plains and the Rockies, these pioneers would have been drawn inexorably south to the rich, dry grasslands and canyons of the Colorado plateau and to the more arid deserts beyond.

While fighting for access to the female herd in the distance, two well-matched Columbian mammoth bulls have accidentally locked their tusks together. Meanwhile, a giant short-faced bear has been drawn to the area by the commotion. In the far distance, a herd of plains bison are on the move in search of fresh grazing.

THE CANYON LANDS

Left: Running for 446 km (277 miles), the Grand Canyon is by far the longest, deepest and most spectacular of the southwest's canyons. **Above centre:** The bizarre-looking Joshua tree of the Mojave Desert. **Below centre:** A desert bighorn sheep, one of the few surviving large mammals of the canyon lands. **Right:** Rarely seen, mountain lions, or cougars, are the biggest predators alive in the southwest today.

Carved by the mighty Colorado River, the awesome Grand Canyon is perhaps North America's best-known geographical landmark.

THE WILD WEST

The American southwest is a vast area of wild deserts, forests, colourful mesas, canyons and mountains. To many this is the Wild West of classic cowboy movies, a land that can often appear harsh and desolate, yet stunningly beautiful. The region boasts a great variety of landscapes and an equally diverse and rich wildlife.

Much of the southwest is characterized by deserts, from the parched wilderness of Death Valley in the west to the Sonoran Desert, with its bizarre forests of columnar cacti, to the south. The southwest is not, however, just a barren, arid land. The huge expanse of deserts is frequently punctuated by green mountain ranges, where forests cloak the slopes and snow blankets the peaks in winter. Many of these mountains exceed 3600 m (12,000 feet) – a stark contrast to Death Valley, which at 86 m (282 feet) below sea level is at the lowest elevation in North America.

To the north and east of the deserts lies the Colorado Plateau, a raised platform of rock dissected by spectacular red rock canyons. The plateau covers some 400,000 sq km (154,440 sq miles) and parts of four states: Arizona, Utah, Colorado and New Mexico. This land of colour and canyons is home to some of North America's best-known and most breathtaking landmarks, including Canyonlands National Park, Arches National Park and, the most famous of them all, the Grand Canyon.

This spectacular gorge, carved by the turbulent waters of the Colorado River, runs for 446 km (277 miles), averages 1.6 km (1 mile) deep and varies from 16 to 29 km (10 to 18 miles) between its northern and southern rims. The diversity of wildlife found within the Canyon is impressive and, in terms of habitats, a hike from the canyon's rim down to the Colorado River is the equivalent of going from the forests of southern Canada to the Sonoran Desert of northern Mexico.

Deep within the canyon, at elevations as low as 300 m (980 feet) above sea level, deserts take hold and heat-loving lizards live among stands of cacti. Nearly 1800 m (5900 feet) above, on the rim, evergreen forests are pock-marked with grassy meadows, where herds of mule deer venture out to graze. Between the rim and the canyon bottom lie hundreds of metres of vertical cliffs, outcrops, rocky slopes and desert flats. This is the domain of desert bighorn sheep and mountain lions.

The canyons, plateaux and mountains of the southwest have a timeless quality, but this conceals the fact that life within them has changed considerably since the region was first explored by humans around 13,000 years ago. So what would have met the eyes of the first people to peer into the depths of the Grand Canyon or at the now-barren deserts of Death Valley?

Left: Mountain lions stalk the canyons for mule deer and bighorn sheep. **Right:** Giant saguaro cacti characterize the Sonoran Desert regions of southern Arizona.

CAVES AND CLUES TO THE PAST

The rugged landscape of the American southwest has an abundance of caves, from vast caverns to tiny rock shelters and crevices. Within these caves, a wonderful treasure trove of fossils provides some excellent clues to the distant past. The arid climate has helped to preserve remains that in other parts of North America would have been degraded by the elements. From the evidence within these caves we can not only learn which animals lived here and what the environment was like during the last Ice Age, but we can unearth fascinating details about these creatures' lives.

How did these fossil remains find their way into the caves in the first place? In some cases the answer is straightforward. The preserved bones belong to animals that once inhabited the caves. Today, as then, caves are used by many species, either as a permanent home or as a temporary base. Some of the larger caves offer occasional shelter to animals such as mountain lions and bighorn sheep, and the small crevices provide homes for rock squirrels, pack-rats and nesting birds. Many species of bat use the caves for roosting and breeding.

But how do we account for the fossils of animals that do not normally use caves? Predators and scavengers would have carried bones back to caves, where they could feed in shelter or feed their offspring. In doing so they inadvertently left clues to be discovered thousands of years later. Other caves acted as natural traps: animals that fell into them were either killed by the fall or were unable to get out and died of starvation.

Regardless of how the animals came to be in the caves, their preservation has helped scientists to piece together a long lost world inhabited by Columbian mammoths, mastodons, camels, horses, Harrington's mountain goats and ground sloths. Living alongside these long-extinct animals were species still alive today such as bighorn sheep, mule deer and mountain lions.

GROUND SLOTHS

Of the wonderful array of beasts present at the end of the last Ice Age perhaps the most bizarre were the ground sloths – strange lumbering creatures that bear little resemblance to anything alive today – which inhabited much of North, Central and South America. Their closest living relatives are the tree sloths of

South America, though there are in fact few similarities between the two. Ground sloths were much bigger than tree sloths, they did not climb trees and nor did they hang upside down.

At the end of the last Ice Age four species of ground sloth lived in North America: Jefferson's, Harlan's, giant and Shasta. Many fossil remains of Jefferson's ground sloth have been found in the southwest but this sloth was common throughout the continent, its range even extending into Beringia in the far north. Uniquely among ground sloth species, the Jefferson's ground sloth – a 2.5 m–3 m (8–10 feet) long beast – walked on the soles, and not on the sides, of its feet. This would have enabled it to move more easily and with greater stability than other ground sloths, and it suited the open grassland environment it lived in. Combined with the animal's greater tolerance of the cold, this adaptation may have allowed Jefferson's ground sloth to colonize areas much further north than the other sloth species.

Though its range did not extend as far north, the Harlan's ground sloth was also widespread in North America and, like Jefferson's, it lived in the southwest. A large creature, up to 3.5 m (11.5 feet) tall, it had a truly bizarre appearance. Its thick leathery skin was reinforced with an outer layer of bony plates, which formed a kind of armour plating. With such an impressive defence, this sloth was unlikely to have had any natural predators.

The giant ground sloth was, as its name implies, a huge beast. It stood approximately 4 m (13 feet) tall and, rearing up on its hind legs, it was almost the height of a giraffe. Like the giraffe, it would have browsed on tree-top leaves well beyond the reach of most other large browsing animals. The giant ground sloth lived in the southeast of the continent in the open forests of Florida (see Chapter 5, p. 151).

The Shasta was the most common of all the ground sloth species in the southwest during the last Ice Age. Clues about its appearance and lifestyle come from New Mexico, where in 1928 a remarkable find was made in a deep pit in Aden Crater. The pit is a volcanic feature known as a fumerole – an opening through which hot gases rise out of the ground. The well-preserved sloth specimen consisted of an entire skeleton as well as ligaments, tendons, muscle fibres, skin, hair and even a ball of dung. The chamber in which the sloth was found lies

Tree sloths, which inhabit the tropical forests of South America, are the closest living relatives of the extinct ground sloths.

some 30 m (98 feet) below the ground and it seems that the sloth had fallen into the pit. Its position when discovered suggests that the sloth was still alive after its initial fall. Near-vertical walls made it impossible for the creature to climb out and instead it ventured down deeper underground passageways. With no way out, the sloth eventually died.

Further insights into this amazing animal have come from Rampart Cave in the Grand Canyon. Not only have bones been discovered here, but also a huge quantity of dried dung, which provides clues to the animal's diet. Sadly a fire in 1976 damaged much of this valuable evidence but, despite this, fascinating information about the sloth's eating habits has been unearthed.

SHASTA GROUND SLOTH

KEY FACTS

Common name: Shasta ground sloth
Scientific name: *Nothrotheriops shastensis*
Size: 2.75 m (9 feet) long
Weight: Up to 250 kg (550 pounds)
Diet: Vegetarian – browsing on a huge variety of different plants
Habitat: Desert scrublands and juniper forests

Ice-Age distribution of Shasta ground sloth

The Shasta was the smallest of the four North American ground sloth species. Nevertheless it was a fairly hefty animal, measuring up to 2.75 m (9 feet) from head to tail and weighing as much as 250 kg (550 pounds). It had large, stout hind legs and by using its powerful muscular tail for support it could rear up to reach food with its heavily clawed front limbs. Although it was not a great athlete, its bulk and long, sharp claws probably discouraged all but the most determined of predators. Sabre-toothed cats may have hunted ground sloths but the risks would have been high. One slash from the sloth's claws could have inflicted a potentially fatal wound. It is thought that the Shasta and other ground sloths defended themselves in much the same way as anteaters do. These distant relatives of the ground sloths are also slow-moving and, at first glance, seem an easy target for hungry predators. When cornered, however, anteaters reveal a very effective defence strategy. Rearing up on their hind legs, they swipe viciously with sharply clawed front legs – an effective deterrent to most would-be predators.

From an examination of the dung at Rampart Cave, it seems that the Shasta ground sloth was not a fussy animal. It certainly had a far less selective diet than any modern animal in the region. It ate a huge variety of plants, including many desert species that can still be found in the area today, such as desert globemallow, cacti and yucca. The sloth's long claws would have been very useful in raking off the fruits from cacti while avoiding their painful spines. It is thought that the Shasta ground sloth had a prehensile tongue, which could wrap around leaves and strip them from branches.

The reason why ground sloths used Rampart Cave remains something of a mystery. There was so much dung in the cave it is clear that sloths made regular use of it. By dating the dung using radiocarbon techniques (see Chapter 1, p. 29) it is known that sloths used the cave from at least 40,000 years ago until around 13,000 years ago, when they became extinct. These caves may have provided shelter from cold winter nights or hot summer days, and offered a more comfortable environment in this land of extremes. In addition, they may have been used as denning sites, where sloths could give birth to their young in relative safety.

Centre: The Shasta ground sloth was quite at home in the arid depths of the Grand Canyon. **Above right:** Cornered, an anteater rears up in a defensive posture. Ground sloths probably adopted a similar strategy when threatened. **Below right:** From analysis of preserved dung in caves, scientists know that the desert globemallow was a favourite food of Shasta ground sloths.

Despite all of these clues, many aspects of the Shasta ground sloth's life are a puzzle. There are no living ground sloths today to provide any insights. The best that scientists can do is to make educated guesses. It seems likely that the Shasta ground sloth was a solitary animal, except of course at breeding times. What little evidence there is suggests that the ground sloth produced a single young. This is based on the find of a similar species in Brazil, *Nothrotherium maquinense*, where the foetus was still preserved in the body cavity. As with tree sloths, a young ground sloth was probably carried by its mother, possibly clinging on to her back. The youngster would have stayed with its mother for several months, until it was large and experienced enough to fend for itself.

Rampart Cave provides evidence of many other animals that once lived in the Ice-Age Grand Canyon. The most common was Harrington's mountain goat, a close relative of the mountain goat that still lives to the north in the Rocky Mountains. Harrington's mountain goat was a stockier animal and well suited to life on the walls of the Grand Canyon. We can tell from the discovery of the preserved hair of this animal that it was white, like the living mountain goat. Shallow depressions in the floor of one particular cave mark where the goats have laid down, presumably to rest during the heat of the day or through the night. By repeatedly lying in the same spot, the goats gradually made impressions of their own bodies and these have survived for over 10,000 years.

VAMPIRES OF THE ICE AGE

One of Rampart Cave's greatest surprises was the discovery of the perfectly preserved skull of an Ice-Age bat. Bats are still common in the Grand Canyon and many kinds emerge at dusk to feed on insects. However, this skull did not belong to an ordinary bat but to an extinct species of vampire bat, the bloodsucker of many a myth and legend. Although it is difficult to put a precise date on this find, it seems the bat lived as recently as 13,000 years ago. Today, other species of vampire bats are found further south in Central and South America, where they feed on the blood of livestock and other large mammals. These bats employ stealth to approach their victims, landing nearby and crawling towards their prey using their wings as limbs. They then inflict a tiny, painless wound with their razor-sharp teeth, and chemicals in the bat's saliva, called anticoagulants, prevent the victim's blood from clotting. The bat can then drink its fill.

During the last Ice Age, the Grand Canyon's larger animals would have provided plenty of food for these vampires. However, given that the living relatives of this bat live in the tropics today, its presence here during the last Ice Age is rather unexpected because of the cooler climate at that time. How the bat became extinct is not clear, but the most likely explanation is that it was due to the disappearance of many of the large animals it fed upon, rather than as a direct consequence of a changing climate.

COLUMBIAN MAMMOTHS

In contrast to vampire bats, mammoths are the animals that perhaps best characterize the Ice Age. Although the woolly mammoth was absent from

Like their extinct cousins, today's mountain goats are well suited to living on exposed, precipitous cliffs.

the relatively warm southern half of the continent, the Columbian mammoth, its close relative, was widespread throughout much of North America (see Chapter 3, p. 86) and there is plenty of evidence that it was common in the southwest. Some of the most interesting clues about these mammoths come from yet another cave – Bechan Cave in Utah. 'Bechan' is derived from a Navajo word meaning 'large dung'. The reason for the cave's name is simple: it is filled with dung, around 300 cubic m (10,600 cubic feet) of it. Most of the dung

Above: Bechan Cave in Utah offers a unique insight into the lives of the Columbian mammoths that once used it for shelter.
Below: The study of preserved dung from Bechan Cave tell us much about the diet of Columbian mammoths.

belonged to the Columbian mammoth and was deposited in large balls, or boluses, very similar to the dung of modern elephants. Analysis of the dung reveals that mammoths were mainly grazing animals and up to 95 per cent of their food came in the form of grass. By dating the dung, scientists have revealed that mammoths were using the cave on and off for up to 1500 years, around 15,000 years ago. Why mammoths were in the cave is not clear, but it could be that they used it for shelter during storms or other periods of extreme weather.

As well as revealing the animals' diets, the dung deposits of Bechan Cave also contained the remains of small dung beetles, belonging to a species known as *Aphodius fossor*, now extinct in North America. As their name implies, these beetles fed on dung, in particular the undigested material contained in herbivore dung. Incredibly, the type of beetle found in mammoth dung is still alive and well in Europe and Asia today. It is a generalist species, which means that it can live off the dung of a wide variety of animals. It appears that the beetle disappeared from the American southwest because of the extinction of mammoths and other large herbivores. However, in Europe and Asia the beetle was able to survive, living off the dung of many different kinds of animal. The role of these beetles cannot be underestimated. They are extremely efficient at recycling animal waste rapidly and, without them, this material would quickly accumulate and smother the land. On the Serengeti plains of Africa, beetles can collectively process as much as 10,000 kg (9.8 tons) of dung each year. During the Ice Age, with so many herbivores present, the role of these beetles would have been vital.

AMERICAN CAMELS

Camels are often regarded as symbolizing the Old World deserts of North Africa and the Middle East. In fact, they first evolved in North America. From there they spread north and then west, crossing the Bering land bridge that linked the American continent to Asia.

The most widespread and best-known species of camel found in Ice-Age North America was the Western or Yesterday's camel. Fossil bone remains of this camel are a common find throughout the American southwest. The Western camel ranged widely across the continent, reaching as far north as Alaska, and it would have been a familiar sight in the open woodlands and grasslands of the southwest. Though similar to the modern dromedary camel, the Western camel was a larger creature, with longer legs, but it is not thought to have had the characteristic hump.

A camel's hump stores fat, which is a valuable energy reserve in desert environments where food may be scarce for long periods of time. However, because the Western camel lived in climates where food was more readily available, this adaptation was probably unnecessary.

The presence of predators such as the sabre-toothed cat meant that there were plenty of carcasses available to scavenging birds and animals during the Ice Age.

CANYON CARNIVORES

The wealth of animal life in the southwest included many meat-eaters. Predators such as wolves, dire wolves, American lions, mountain lions and sabre-toothed cats lived alongside a number of animals that specialized in scavenging. There would have been a constant supply of dead animals, both those that had died of old age, starvation or disease and those that had had died at the hands of hunters. Studies of the teeth of sabre-toothed cats suggest that they fed only on the soft, fleshy parts of their victims. This may have been because their long, relatively narrow teeth were prone to damage if the cat accidentally bit into bone. For a sabre-tooth, damage to its canine teeth could prove fatal. Without their most effective weapons, hunting would have been almost impossible. So it seems likely that, having killed and fed, a sabre-tooth would have left a lot of meat on the carcass – a source of food that was readily exploited by the scavengers of the Ice Age.

AERIAL SCAVENGERS

Among the scavengers there was a range of bird species that specialized in eating dead animals. Once again, caves provide much of the evidence for these aerial scavengers. Today, turkey vultures and ravens nest on the high walls, ledges and caves of the Grand Canyon. They live largely by scavenging on the carcasses of dead animals and on the scraps of food left behind by human visitors.

Thirteen thousand years ago there was a much bigger scavenger here – the California condor, a species that still maintains a tenuous grip on survival. In Sandblast Cave, in the Grand Canyon, the remains of condors have been found, along with fragments of eggshell and feathers. The bones appear to have come from young condors yet to reach fledgling age, suggesting that the cave was used as a nesting site. Other animal remains were also found in Sandblast Cave, including the bones of horses, camels, bison and mammoths. There is no way these animals could have reached the cave by themselves – it is too inaccessible. The small size of the bones suggests that they were carried there by adult condors as food for their young. With a wingspan of over 2.5 m (8.2 feet), condors are

REINTRODUCING CONDORS TO THE GRAND CANYON

The California condor died out from the Grand Canyon over 11,000 years ago. It nearly became extinct altogether, but recently a programme of captive breeding has saved this ancient scavenger and after a long absence it can once again be seen soaring over the Grand Canyon.

By the mid-1980s the last wild condors could only be found in a very small area of California. In a desperate attempt to save the species, this handful of remaining birds was caught and brought into captivity. A successful breeding programme enabled a number of young condors to be reared safely, free from the risks that threaten them in the wild. Once a sufficient number had been bred, a reintroduction scheme was initiated.

At first, the birds were released back into areas where they had last flown in the wild. Following this success, it was decided to start a new wild population. A release site was chosen along the Vermillion Cliffs wilderness area, 48 km (30 miles) north of the Grand Canyon, and in 1996 six birds were released. More have been introduced since to bolster the small founding group.

Condors can cover vast distances with effortless ease because of their huge wingspans. In no time at all, the condors began to explore the surrounding country and they are now regular visitors to the Grand Canyon, sometimes even putting on a show for the throngs of visitors along the south rim.

Vultures and a maribou stork gather at a carcass on the African plains. Scenes like this would once have been commonplace in the American southwest.

the biggest living land birds in North America. Their long, broad wings enable them to glide and soar, and they can cover considerable distances in search of food. They would have used the updraughts from the walls of the Grand Canyon to gain height before setting off in search of dead animals on the surrounding plateau. Having found a carcass, the adult condors would gorge themselves before carrying what scraps they could back to their youngsters in the nest cave.

California condors were not the only birds to spend their lives scouring the land for dead animals during the last Ice Age. In fact, evidence from elsewhere in the western United States points to the existence of a whole range of scavenging birds, each carving out its own particular niche. There was another condor – the La Brea condor – and a huge condor-like bird called the teratorn (see Chapter 2, p. 59). In addition, there was a variety of vultures, including the turkey and black vultures (still present in the region today) and vultures more closely related to those now found in Europe, Africa and Asia. Eagles and caracaras add to the list of scavenging birds, and there was even a stork, which may have lived like the marabou stork of Africa, a regular attendant at any dead animal feast.

The scavenging birds of Ice-Age North America were comparable to those found on the plains of Africa today. Here, a variety of vultures, eagles and storks will gather around a dead animal, with each bird adapted to make use of the carcass in its own unique way. The first to feed are the larger vultures; their powerful beaks enable them to rip open the carcass and tear off large pieces of tough hide. Medium-sized vultures follow; they tend to go for the softer, fleshier parts that the big vultures have exposed. On the sidelines, smaller vultures and other scavenging birds wait, looking for opportunities to sneak in and grab small scraps.

In North America, most of these birds are now extinct, presumably because their existence depended upon a bountiful supply of large dead animals. When the megafauna declined, the scavengers were unable to find alternative food sources. Only the most adaptable species, such as turkey and black vultures, survived. Less specialized than their extinct relatives, it was their broader tastes that saved them.

THE TROUBLE WITH BONES

Most of the evidence for the animals that inhabited North America during the Ice Age comes in the form of bones. Occasionally a complete or near-complete skeleton is unearthed, but most of the time scientists have to rely on their ability to identify individual bones or even fragments of bones. Assigning bones to individual species is fraught with difficulty, particularly with smaller animals. Birds provide a good example: most bird bones can only be identified to the level of a group of species. It might be possible to say that a leg bone comes from a grouse or a shorebird, but that is as close as it gets. Even larger birds pose problems and there is often disagreement between scientists as to how many species really existed.

The same can be said of other extinct animals. Often a single specimen, maybe just one or two bones, is all that has been discovered for an entire species. As a result, scientists are forever re-evaluating how many species there really were and what they looked like. New techniques are constantly developing, and as new finds are made and old ones are re-investigated, the picture of Ice-Age North America is ever changing.

ICE-AGE LANDSCAPES

From the evidence collected so far, we can imagine what the Grand Canyon must have been like around 13,000 years ago. Along the rim, herds of horses, bison and mammoths grazed in grassy forest clearings, while camels browsed on trees and shrubs. Far below, on the precipitous canyon walls, Harrington's mountain goats leaped from ledge to ledge in search of succulent plants well beyond the reach of less athletic animals. Lower still, Shasta ground sloths moved slowly across the inner plateaux, browsing on a variety of plants. Circling overhead, condors searched for a meal – perhaps a mountain goat that had missed its footing and plunged to its death.

This abundance of animals is hard to imagine when visiting the region today. Visitors may spot the occasional mule deer, but the landscape is relatively parched and it is difficult to envisage a world inhabited by herds of big animals. So what enabled these creatures to flourish during the last Ice Age? The answer lies in the vastly different environment that prevailed in the southwest at the time. The Grand Canyon may still have existed – indeed it has done so for the last 6 million years – but the vegetation that clothed it was quite different.

Left: The Grand Canyon today is fairly arid and is home to only a handful of large animals. **Centre:** The wild burro is descended from wild African asses, but 13,000 years ago the region had its own similar species. **Right:** Mule deer are one of the few large mammals that visitors to the Grand Canyon are likely to spot today.

So how can we build a picture of the habitat 13,000 years ago? The animals themselves provide some clues. Mammoths, horses and bison are all grazers, therefore there must have been plenty of grass to support these beasts. The Shasta ground sloth's dung gives direct evidence of all sorts of plants, including many desert species that are still found today. However, this evidence alone cannot help reconstruct past habitats. There may have been grass, but was it found in dry desert grasslands, open prairies, grassy forest clearings or open meadows?

THE WILD BURRO

When Europeans arrived in North America they brought many different animals with them. For example, a domesticated form of the African wild ass, the burro, was introduced by the Spanish in the 1500s. These small horses fared well and have effectively become wild animals in the southwest. The fact that they have adapted so readily is perhaps not surprising.

Up until the end of the Ice Age the region had its own native burro-type horse. This animal was similar in size and build to the modern burro. With its introduction to the southwest, the burro was able to carry on where its extinct cousin had left off. Given that a horse similar to the burro once lived in the southwest, it is perhaps ironic that today the burro is considered by many to be a pest.

Burros are well adapted to the environment and out-compete many native animals for food and water resources. A reduction in the number of animals such as bighorn sheep could be attributed to the success of burros. As a result, programmes are in place to control their populations.

Left: Desert pack-rats are avid collectors and their efforts have enabled scientists to reconstruct past environments.
Right: A preserved pack-rat midden holds clues to landscapes dating back tens of thousands of years.

THE PACK-RAT'S TALE

Much of what we know about the Ice-Age landscape comes from a very unlikely source – an industrious little rodent called the pack-rat. There are several kinds of pack-rat, or wood-rat, and many of them share a kleptomania habit. Pack-rats build nests, often tucked into the shelter of a rock crevice or in a cave. They gather all sorts of objects into their nest – sticks, seeds, feathers, insects, pieces of bone – which add to the protective structure. Some nests have even been found to contain false teeth, silverware and cigarette lighters. Outside its nest, the pack-rat deposits various bits of this material onto a kind of garbage pile or midden, upon which it often urinates. The pack-rat's urine is extremely viscous and binds the debris together. The midden gradually hardens and solidifies into a rock-like protective barrier outside the nest. The crystallized pack-rat urine is known as amberat. When early European miners first discovered this amberat they thought it was some kind of sweet and referred to it as 'rock-candy'. The amberat was far from sweet and caused a few rather unpleasant bouts of nausea!

Even though pack-rats only live for a few years, successive generations will often use the same midden, which, in time, can become very large. When the middens are built into dry rock crevices or caves, where little or no moisture reaches them, they can persist for literally thousands of years. These ancient pack-rat middens are referred to as paleo-middens. Some have even been dated to around 45,000 years old, which is the effective limit of radiocarbon dating. Others may, therefore, be even older. The paleo-middens contain a perfectly preserved sample of the vegetation from the surrounding area. Seeds, leaves, pine needles, pieces of wood and pollen can all be examined, allowing scientists to reconstruct what the area might have looked like – often with surprising results.

It has been established that during the last Ice Age the deserts and the saguaro cactus forests of the southwest did not exist as we know them today. Instead, much of the area was blanketed in forests of pinyon pine, juniper and oak. As a rule, the plants we see today lived 600–1000 m (1970–3280 feet) lower in elevation. Desert plants still existed, but in an unusual mixture with other plants and trees. Barrel cacti lived among pine trees and, in the arid depths of the Grand Canyon, juniper bushes thrived alongside cholla cacti. At higher elevations, the landscape was much greener than is found today. On the Colorado Plateau, open forests of fir and spruce thrived, interspersed with grassland areas, creating what scientists refer to as a 'parkland' environment. People arriving in the Grand Canyon 13,000 years ago would have been confronted with a much greener and lusher view than visitors encounter today.

Other forms of evidence help to support this theory. The study of pollen and other plant remains, laid down thousands of years ago and preserved in lake sediments, has led to the same conclusions. The southwest was not home to the deserts that now characterize much of the region. Only a small area of what can be called true desert was found, in the lower reaches of the Colorado River valley, an area that is extremely hot and arid today. As the Ice Age finally released its grip on the continent around 13,000 years ago, the weather patterns started to change and the deserts as we now recognize them slowly began to take shape.

A kind of large yucca, Joshua trees grow in the Mojave Desert of southern California.

THE SOUTHWEST DESERTS TODAY

All of the major deserts in North America fall within the southwest. They are subdivided into four general types, each with its own unique characteristics – the Mojave, the Sonoran, the Chihuahuan and the Great Basin Deserts. By definition, all deserts share an arid climate, typically receiving less than 250 mm (10 inches) of precipitation a year. Substantial amounts of this moisture is lost through evaporation, with little available to living creatures and plants. While the deserts of the southwest have this aridity in common, they do differ in many distinctive ways.

The Great Basin Desert is the largest desert region within the United States. It covers an area of approximately 490,000 sq km (190,000 sq miles) and is bordered to the west by the Sierra Nevada Mountains and to the east by the Rockies. The Great Basin is a relatively cool desert due to its more northerly latitude and relatively high elevation, with much of the desert at 1200–1800 m (4000–6000 feet). As a consequence, precipitation often falls as snow through the winter months. There is not a great diversity of plant life, and just a few low-growing shrubs dominate, including plants like sagebrush and Mormon-tea.

The Mojave Desert is sandwiched between the Great Basin Desert to the north and the Sonoran Desert to the south. The Mojave is very arid, receiving less than 125 mm (5 inches) of precipitation a year, most of which falls during the winter. It is a land of extremes, with sometimes freezing winters and hot, dry and windy summers. Lower parts of this desert can be almost devoid of plant life. The barren landscape of Death Valley falls within the Mojave Desert and its name reflects the harshness of the land – one of the hottest and driest places on Earth. In the higher elevations of the Mojave, a strange tree-like yucca called the Joshua tree can be found. Entire 'forests' of Joshua trees occur in some areas, creating a truly distinctive and unique landscape.

The Chihuahuan Desert is the largest of the North American deserts, if Mexico is included within this geographical description. It covers over 520,000 sq km (200,800 sq miles) but most of this area lies south of the Mexican border. It is a desert characterized by shrubs and it contains relatively little variety of plant life. Yuccas and agaves, growing amongst grasses, and creosote bushes are typical of this landscape.

Left: Growing up to 9 m (30 feet) tall, saguaro cacti can live for over 200 years. **Right:** Many species of cactus produce spectacular blooms, like this hedgehog cactus.

The Sonoran Desert covers an area of about 311,000 sq km (120,000 sq miles). In the United States it occupies a sizeable part of southwestern Arizona and southeastern California; from here it stretches south into Mexico. It is the hottest desert in North America, with summer temperatures regularly exceeding 40ºC (104ºF). In one area of the Sonoran Desert, known as the Lower Colorado Desert, temperatures can rise even higher – close to 50ºC (122ºF) – making it the hottest place in North America. In spite of this heat and aridity, unique rainfall patterns create a diversity of life not found in the other deserts. Winter storms can induce spectacular blooms of colourful wild flowers, and summer monsoons encourage both annuals and woody plants to flourish. Trees and shrubs are diverse, but the most characteristic plant is surely the saguaro cactus. This columnar cactus dominates many parts of the Sonoran Desert and can grow up to 9 m (30 feet) tall. The oldest plants are estimated to be over 200 years old.

DESERT LIFE

For the animals and plants living in North America's deserts, the biggest problem is coping with the heat and aridity. Avoiding the heat is one of the most important factors dictating the behaviour of many desert animals. Many limit their activities above ground to the hours of darkness to avoid the heat of the day. A great number of birds leave the deserts during the hottest time of the year, migrating to more hospitable climates. Spadefoot toads spend much of the year entombed in burrows, where they can remain cool and moist. Only when the rains arrive do they dig their way to the surface to mate and breed in temporary pools.

Many animals have specific adaptations to enable them to cope with the heat. For example, jackrabbits in desert areas have huge ears, with a dense network of blood vessels, which help the animals to lose heat quickly. Turkey and black vultures lose heat by urinating onto their legs, which then cool by evaporation. The cooled blood is circulated back through the body, helping to maintain a stable temperature.

Avoiding water loss is another important adaptation to desert life. Birds and reptiles excrete insoluble uric acid, wasting very little precious water in the process. However, mammals excrete in the form of urine and they lose valuable water as a result. For most mammals, therefore, access to a good water supply is essential, though there are exceptions.

A DESERT SPECIALIST

Kangaroo rats are remarkable animals and true desert specialists. They derive their name from their disproportionately large back legs and feet, which enable them to hop along with remarkable agility. A long tail provides an effective counterbalance.

Kangaroo rats spend a lot of their time in underground burrows, which they block off during the day to keep out the heat. They do not sweat or pant to keep cool, as other animals do, and so they avoid valuable moisture loss. Specialized kidneys help to extract most of the water from their urine, returning it to the bloodstream. Water that would normally be lost through exhaling is recaptured by special organs in the nasal cavities. The kangaroo rat can even make its own water through the digestion of dry seeds. So adept is it at life in the deserts that it does not need to drink. Indeed, if offered water in captivity, it refuses to drink.

COOLER, WETTER, GREENER

Why were the habitats that prevailed in the Ice Age so different from those we see today? The simple answer is that the climate has changed. In broad terms, the region generally experienced a more equable climate during the Ice Age, without the extremes of today. It was, on average, about 5°C (9°F) cooler. Rainfall was probably higher too, anywhere from between 35 and 65 per cent higher than present-day levels. More significantly, the seasonal patterns of rainfall were different. Winters were wetter and summers were drier, which in turn had a dramatic effect on the region's vegetation. Ponderosa pine provides a good example and illustrates how these climatic differences were not as straightfoward as one might imagine.

Ponderosa is a common and widespread tree on the Colorado Plateau, but it is unexpectedly absent from the fossil record. Variations in temperature alone do not account for its absence, but differences in rainfall patterns do. Ponderosa can only grow if it receives sufficient summer rainfall. During the Ice Age the climate may have been wetter on average, but the rainfall was concentrated in the winter, and summers were often dry – too dry for ponderosa.

Pinyon pine is another common tree on the plateau today, but it too was virtually non-existent during the Ice Age. This may be partly due to lower temperatures: on the rim of the Grand Canyon temperatures would have regularly fallen below those conducive to the growth of this species. Nevertheless, colder conditions alone do not account for the lack of pinyon. If they did, the pine would have been found lower down the canyon, where temperatures were higher. The scarcity of pinyon is probably explained by the combined effect of a dry climate and lower temperatures. Today, pinyon lives in areas as dry as the Ice-Age Grand Canyon, and in areas as cool, but it does not live where it is both as dry and as cool.

These examples show how complicated the discrepancies are between today's environment and the one that existed at the end of the last Ice Age. It is tempting to assume that the Ice Age was cooler and wetter and that the plants simply shifted their distribution by moving south or downwards in elevation. The examples of ponderosa and pinyon pines illustrate the fact that the differences were far more subtle.

Left: As the Ice Age ended, forests of pinyon and juniper spread northwards; they now cover much of the southwest. **Right:** Ponderosa pines are able to grow in even the most precarious of locations.

THE PINYON PINE STORY

After the Ice Age pinyon pine made a rapid advance northwards. It could do this because of the work of birds living in pinyon forests.

Pinyon has particularly nutritious seeds, which appeal to a range of creatures, among them the pinyon jay and Clark's nutcracker. When pinyon seeds are in plentiful supply these birds will store them to eat later on during leaner times. They gather up seeds in their crop, a special sac-like structure in the throat, and fly off to bury them in the soil, often a long distance away.

Studies have shown that both the jay and the nutcracker have a remarkable memory for relocating buried seeds. But neither is infallible. Many seeds are never found and are left undisturbed to germinate, allowing the pinyon forests to spread.

Salt flats are the dried-up beds of the ancient lakes that dotted the southwest during the Ice Age.

LAKES AND WETLANDS

One of the most striking contrasts between the southwest today and at the end of the Ice Age was the presence then of numerous large shallow lakes, which occupied 'closed basins'. This simply means that the lakes had no outflow of water. Called pluvial lakes, these bodies of water existed because of the higher rainfall and cooler climate, which meant there was less evaporation. Melting glaciers in the region's mountain ranges may have helped to maintain the lakes, which even occurred in some of the world's most inhospitable places, such as Death Valley.

Salt Lake in Utah is a relic of one of these pluvial lakes – Lake Bonneville – which, with a surface area of about 51,300 sq km (19,800 sq miles), once covered much of western Utah. The Bonneville salt flats are the hardened, sun-baked bed of this ancient lake. To the west of Bonneville, in Nevada, the Black Rock Desert is a similar location. The desert flats here are the ancient bed of Lake Lahontan, which at its greatest extent covered about 22,400 sq km (8650 sq miles) and had a maximum depth of 270 m (885 feet).

Towards the end of the Ice Age, Lake Bonneville flooded. The Bonneville floods are thought to have lasted from between two months and one year; they were second only in scale to the Lake Missoula floods (see Chapter 2, p. 49). If, as some argue, people did arrive in North America before 13,000 years ago, perhaps they witnessed this natural catastrophe and became victims of the flood.

The wildlife of these lakes and their associated wetlands would have been impressive. The lakes would have been like oases to large animals such as mammoths, horses and camels. Waterbirds would have thrived, some breeding in the wetlands and others using the lakes as resting areas on their migrations to and from breeding grounds further north. It is even possible that birds such as flamingos lived here. Evidence from California suggests that a species of flamingo living in North America at the end of the Ice Age may be the same as the greater flamingo that now inhabits the Caribbean. It seems likely that this flamingo bred on the shallow lakes of the Great Basin and Mojave Deserts, feeding in the plankton-rich waters.

As the climate changed towards the end of the Ice Age, these lakes began to dry up. With higher temperatures, evaporation from the lakes began to increase. Rainfall declined and the flow of melt waters from the ever-shrinking glaciers decreased. As a result, the water in the pluvial lakes was not replaced quickly enough and levels dropped. In time, many lakes dried up altogether, leaving the salt flats we see today. Only the largest lakes survived, albeit in much smaller forms.

So, as we have shown, this region has undergone dramatic changes in its wildlife and landscape. These changes have been relatively swift and dramatic. Around 13,000 years ago, when the first people arrived on the Colorado Plateau, in Death Valley and in the Sonoran Desert, they found a much greener world with an abundance of animal life. These first North Americans shared the land not only with the wildlife we can see today, but with a whole host of other strange and wonderful animals.

Many of these animals lived in North America's most southerly extreme, Florida, where the first people also settled. Florida and the rest of the Deep South was a warm refuge during the Ice Age, containing perhaps the richest and most diverse wildlife in the continent. It is here that our investigation continues.

A scene from the American southwest around 13,000 years ago: Columbian mammoths and horses graze on the dry grasslands as a hungry sabre-toothed cat looks on, sizing up a potential meal.

FLORIDA: ICE-AGE OASIS

Left: Florida today has an almost tropical feel and seems a world away from the Ice Age. **Above centre:** The jaguar was one of the top predators in the area. **Below centre:** Living on islands of sand, beach mice are perfectly colour-matched to their habitat. **Right:** The beaver is the largest rodent in North America, but it is small compared with the Ice-Age giant beaver, which was the size of a bear.

A TASTE OF THE TROPICS

Jutting out from the southeast corner of the North American continent is a huge low-lying peninsula that stretches 640 km (400 miles) into the Gulf of Mexico towards the warm waters of the Caribbean. Aptly named 'the sunshine state', Florida is so far south it feels more like the tropics than a part of North America. The sun shines for an average of eight hours a day, and every year the land is drenched with over 150 cm (60 inches) of rain. This combination of sun and rain results in a profusion of lush vegetation that would not look out of place in an equatorial rainforest: a dazzling array of palms, ferns, orchids and trees dripping with Spanish moss. More species of trees, reptiles, amphibians and fish live here than anywhere else in the continent.

The key to much of Florida's abundant wildlife is water. This is one of the wettest areas in the continent – one-fifth of the land is under water. From the air, countless lakes, ponds, sinkholes, rivers and creeks are visible. Surrounded by almost 13,500 km (8400 miles) of tidal shoreline, nowhere in the state is further than 100 km (62 miles) from the ocean. The land barely rises above the water that surrounds it, and at its highest point is a mere 100 m (328 feet) above sea level.

The boundary between land and water is often difficult to define, and this is where vast mangrove swamps and marshes form. The most famous marsh of all is the Everglades. Spanning 1 million hectares (2.5 million acres), this subtropical national park covers most of southern Florida. Flooded saw-grass prairie stretches as far as the eye can see, broken only by the occasional hammock of trees.

The favourable conditions that make Florida so appealing to wildlife hold a similar attraction for humans. The state is now home to some 16 million residents, and 1000 new settlers arrive every day. Moreover, Florida is one of the world's most popular holiday destinations, attracting 78 million visitors a year. People are, however, only comparatively recent arrivals. Thirteen thousand years they ago were exploring this land for the very first time. Florida then was a very different place. The land was dominated not by humans but by wildlife, including some of the most extraordinary creatures that have ever lived. Today the

A great blue heron looks out across the Everglades National Park, the largest remaining subtropical wilderness in North America.

state seems a world away from the Ice Age, and even at the peak of glaciation around 20,000 years ago the vast ice sheets were thousands of kilometres to the north. Nevertheless, it was this part of the continent that saw one of the most dramatic consequences of the ice.

So what was Florida like during the Ice Age? What sights met the first colonizers of the southeast, and which beasts roamed the landscape all those thousands of years ago? Florida has one of the richest and most comprehensive Pleistocene fossil records in the world. In the rest of the continent, scientists have looked to the land itself to travel back in time and investigate Ice-Age North America. They have discovered clues to the past in the fossils and frozen animals that have been found in caves, tar pits and permafrost. To uncover Florida's fossil records you have to dive underwater and search one of the natural wonders of the southeast – freshwater springs.

Florida's surface is dotted with hundreds of springs. Fern Hammock Springs is seen here in the morning mist.

FLOODED GRAVEYARDS

Geologically speaking, Florida is a baby. For most of its history it has been submerged under the ocean. While dinosaurs were roaming the rest of the North American continent, Florida was underneath a shallow tropical sea that teemed with life. Giant sharks and a profusion of sea creatures swam above huge coral reefs. Over millions of years, the fossilized remains of these creatures formed massive beds of limestone rock, which in places extend over 3000 m (9,800 feet) deep.

Twenty-five million years ago, Florida emerged from the oceans and was gradually covered in lush vegetation. Rainwater that fell on rotting leaves carried plant acids into the ground, slowly dissolving the limestone. Over time, tunnels and cracks formed. These were small at first, but as more water found its way into the limestone, they grew bigger. After millions of years the result is an incredible honeycomb network of interconnecting tunnels, passageways and caverns. Now flooded with rainwater, this limestone forms the underground water table, or aquifer. Florida is essentially a vast rock sponge, collecting rainwater from as far away as the Appalachian Mountains.

In many parts of Florida, this rock sponge is very close to ground level and, when combined with pressure from within the system, water is forced to the surface as a spring. Northern Florida in particular is covered with these springs,

SPACE FOR WILDLIFE

Thirteen thousand years after people first set foot in this region, our unquenchable desire to explore and colonize has been taken to astronomical limits. Since 1953, Cape Canaveral on Florida's east coast has been the world's most important site for rocket and shuttle launches into space.

Yet even amid this very human activity, wildlife flourishes. Under the shadow of the huge NASA buildings, Merritt Island National Wildlife Refuge covers 56,000 hectares (138,000 acres). With many areas inaccessible to the public, animals are able to live undisturbed. There are 330 species of birds, 31 of mammals, 117 of fish, and 65 of amphibians and reptiles. Bald eagles nest within sight of the NASA flag, ospreys fly past the huge space vehicles to fish in the creeks, and manatees feed in the reflection of the shuttle launches. The incredible noise from the shuttle's booster rockets causes alligators to bellow as they mistake the din for the call of another territorial male.

some small, others huge. The bowl of Wakulla Springs, near Tallahassee, drops down over 55 m (180 feet) below the surface and stretches 75 m (246 feet) across. Silver Springs, north of Orlando, is the world's largest spring, pouring forth around 2 billion litres (0.5 billion gallons) of water every day – enough to fill an Olympic swimming pool every one and a half minutes.

Fresh water can be dark, murky and cloudy, but these springs are crystal-clear exceptions, free from sediment or discoloration. Visibility in many of them is more akin to a tropical coral sea. When viewed in sunlight from above, they are an almost unnatural blue, like the seas around a tropical island. The waters are also relatively warm: coming from deep below the Earth's surface, they emerge at a constant 22°C (72°F).

On cooler mornings, steam rises from the water, shrouding the springs in a fine mist. As well as being beautiful, the springs have an air of mystery for they hold the secrets of the past. The bottom of many of Florida's springs and sinkholes are littered with fossils. When divers first entered Florida's waterways, they were confronted by graveyards of bones, which in some places were literally piled on top of each other. Today, collectors have removed many of the larger fossils, but a visitor to any of Florida's spring-fed rivers will still see small fragments of bone and teeth.

THE FIRST FLORIDIANS

In the 1930s, a diver investigating Ichetuckneee Springs, north of Gainesville, found the partly articulated skeleton of a mastodon. The fact that the skeleton was articulated meant that it had not been disturbed. Beneath one of the vertebrae the diver made a remarkable discovery – a flint scraper. This was the first time in Florida that Paleo-Indian artefacts had been found underwater in association with the remains of extinct animals. It was the start of the underwater search for evidence of the first Floridians. To date, that search has uncovered some of the most important finds in North America. It gives unequivocal evidence that people did indeed meet with the Ice-Age megafauna. Arguably, the most important location for these finds is the Aucilla River, south of Florida's capital city, Tallahassee.

The Aucilla River is a dark and mysterious place with thick woodland covering its banks. The river flows almost imperceptibly, meandering through the eerie

The dark, mysterious waters of the Aucilla River hold many secrets to Florida's past.

THE DATING GAME

Obtaining unequivocal dates for the earliest people is, in reality, an extremely difficult task. The organic bone remains of animals can be carbon dated, but carbon dating does not work on flint or stone artefacts such as spear points. Therefore, it is difficult to say with absolute certainty how old many of these flint points are. It could be argued that stone tools found in the same sediment as fossil skeletons might have become mixed up at a later date.

Occasionally, incontrovertible evidence comes to light. In 1980, a bison skull was found in the Wacissa River in northern Florida. This did not belong to a modern bison but to the Ice-Age long-horned bison. At the top of the skull was an obscure patch of material that was clearly not bone. When the skull was examined under a CAT scan, it was possible to make out the unmistakable shape of a flint spear point, lodged in the bone. This proved that the bison had been killed by people. Carbon dates from the bone gives an accurate date of about 13,000 years ago.

Some artefacts, however, provide evidence of an early human presence without even the need to carry out carbon dating. Tools have been found made from the bones of extinct animals, for example bone daggers made from horse leg bones. Since we know that horses died out from North America at the end of the last Ice Age 12,000 years ago, we can deduce that people must have been present before that date.

landscape for a few hundred metres before disappearing under a solid wall of rock, only to reappear a little way downstream through another sinkhole. In sharp contrast to the crystal-clear waters of many of Florida's spring-fed waterways, the Aucilla is cloudy and stained with tannin. Sunlight rarely reaches the river bed; its secrets are not on view for all to see.

Since 1983, teams of scientists and volunteers have been diving into the gloomy depths of the Aucilla River and slowly excavating the sediments at the river bottom. Amazingly, the river's very gentle current has meant that the layers of sediment lie undisturbed. Even the lightest of plant material, pollen, has settled and been recovered. Buried here is a virtually complete record of human settlement from the present day back through thousands of years to the Paleo-Indian period.

In 1993, an important discovery was made when a 2 m (6.5 feet) long mastodon ivory tusk was found in a layer of sediment dated at over 12,200 years old. When it was brought to the surface, the tusk had eight long and distinctive cut marks near its base, where it had been butchered and removed from the skull by humans. Ivory foreshafts have also been found here – slender, carved shafts of mastodon tusk that were part of a Paleo-Indian spear. Some of the foreshafts are almost 30 cm (12 inches) long. In the middle of one foreshaft is a remarkable geometric, zig-zag engraving – the earliest known artwork in North America.

Further insights into the lives of the Paleo-Indians have come from the discovery of the Ryan-Harley site on the Wacissa River. Here, submerged in less than 1 m (3.3 feet) of water, are the remains of a Paleo-Indian settlement. Again, flint points have been found in connection with extinct megafauna. What is perhaps most interesting is the glimpse of another side to Paleo-Indian life. We now know that these people were skilled and accomplished hunters of mastodons and other large mammals (see Chapter 6, p. 168). Here there is evidence of a much wider use of other wildlife. The fossils suggest that the first people were catching smaller animals such as mink, rabbits and turtles. It also appears that they were proficient anglers. Bone fish-hooks made from raccoon bacula (penis bone), bone spear points and many fish remains were all found at this site. Life for the first people in Florida must have seemed good. They had all they needed: a supply of fresh water, access to chert – the raw material for making flint tools – and above all lots of animals to hunt.

CURIOUS CREATURES

What is immediately apparent from looking at the fossils is that Florida's Ice-Age wildlife was incredibly rich and varied. Many fossils are easily recognizable because they are from animals that still live in the state today – white-tailed deer, panthers, raccoons, beavers and alligators. There are also many extinct species, such as horses, llamas, bison and peccaries. And then there are the Ice-Age giants – the mastodons and Colombian mammoths.

One extinct Ice-Age animal in particular had a truly odd appearance. Curious rosette-shaped fossils, up to a few centimetres across and with regular hexagonal patterning, have been discovered in Florida's waterways. Like a jigsaw, it is only when they are pieced together that the bizarre creature they come from becomes visible. These rosettes are actually dermal scutes, or scales, belonging to a glyptodont.

MANATEES-THE MAMMOTHS' SURVIVORS

Although superficially resembling seals, West Indian manatees are quite unlike any other North American animal. Their closest living relative is in fact the elephant, and manatees are the only surviving relatives in North America of the mammoths that once roamed the continent. They have no hind legs; instead they have a single large, round flipper for a tail. Their fore-limbs are seal-like and form large, surprisingly dextrous flippers, each with three to four claws. These flippers are used for 'walking' on the river bed and for handling aquatic vegetation.

Manatees grow up to 3.5 m (11.5 feet) long and weigh up to 1000 kg (0.9 tons). In areas where they are not hunted or disturbed, they can become very tame. These gentle giants typically spend their day moving around shallow estuaries, rivers and lagoons, where they feed on numerous aquatic plants, sea grass being a favourite. A single animal will consume 10–15 per cent of its body weight a day – as much as 150 kg (330 pounds).

Even though manatees grow to an immense size, they do not have a thick layer of fat or blubber, and their metabolic rate is surprisingly low. As a result, they lose heat rapidly to the surrounding water and so only live in relatively warm water. Manatees are mainly found in the Caribbean. Florida may have a sub-tropical climate, but manatees only just survive here and during winter the whole population is forced to migrate to the state's warm springs. If it were not for these spring refuges, manatees would not be able to exist here. It is unlikely that manatees were present in the continent during the last Ice Age because of the cooler climate at that time.

GLYPTODONT

KEY FACTS
Common name: Glyptodont
Scientific name: *Glypototherium floridanum*
Size: 1.5 m (4.9 feet) tall; 3 m (9.8 feet) long
Weight: 1000 kg (0.9 tons)
Diet: Vegetarian
Habitat: Grasslands and water's edge

Many of the extinct animals of North America have living counterparts that we can turn to for ideas on how they might have looked, lived and moved. For mammoths and mastodons, we can look to elephants; for scimitar- and sabre-toothed cats, today's big cats provide a guide. But the glyptodont is like nothing on Earth.

This animal is so extraordinary that nineteenth-century scientists were baffled when trying to classify it. Glyptodonts are placed in the order of edentates, which is made up of a primitive group of mammals whose teeth have no enamel. They share this order with sloths and armadillos. But although resembling an oversized armadillo, closer inspection of the glyptodont reveals a mix of other animals. It has a unique skull with modifications not found in any other mammal. The glyptodont's teeth resemble a rodent's and are quite unlike an armadillo's; much of its vertebrae and pelvis are fused in the same way a tortoise; and some scientists believe that it may have had a flexible proboscis-like snout, similar to a tapir's. Most extraordinary of all was its shell or carapace.

Virtually all of the glyptodont's body was covered in bony scutes. As many as 1800 scutes were fused tightly together to form a rigid, impenetrable armour up to 5 cm (2 inches) thick. Each of the scutes was pitted with holes, through which hairs protruded from the skin. Its head was also covered in a tough shield. Glyptodonts grew to a considerable size – up to 1.5 m (4.9 feet) tall, over 3 m (9.8 feet) long from nose to tail – and weighed an estimated 1000 kg (0.9 tons), or as much as an average-sized car. In North America, there were five species, all very similar in appearance and all restricted to the southeastern parts of the continent. In South America there were many more species, some of which grew even larger.

The name 'glyptodont' comes from the Greek for 'carved [or grooved] tooth'. By examining glyptodont teeth, scientists have deduced much about how the animal lived. Teeth in all animals have evolved specifically to deal with particular diets, and unrelated animals with similar diets have similar teeth. Glyptodonts have unusual teeth for such a large mammal. They closely resemble the teeth of capybaras – giant rodents that look like huge, dog-sized guinea pigs. One species of capybara is still alive today in the flooded grasslands of South America, where it feeds almost exclusively on emergent aquatic vegetation. Its teeth and jaw muscles are designed to chew relatively tough plant fibres. From the similarity of their teeth, it seems likely that glyptodonts also fed on waterside vegetation. The common occurrence of capybara fossils alongside glyptodonts supports this theory and suggests that they once shared the same habitat – probably the edge of watercourses with shallow and dense vegetation.

The glyptodont's legs were stocky, immensely strong and designed for supporting its heavy weight. Five hoof-like toes on

Above: Now found only in South America, tapir were once common throughout Florida. Glyptodonts may have had a similar proboscis-like snout. **Below:** The massive glyptodont was covered in a thick, almost impenetrable, armour.

each foot spread its weight across the ground it walked on. The movement of its back legs would have been restricted, as the glyptodont's pelvic bones were fused to its shell. Its long tail was covered in a series of interlocking armoured rings, which made it flexible and probably acted as a counter-balance when the animal walked.

Glyptodonts would not have been capable of fast movement and doubtless relied on their armour to defend themselves against predators. The only part of a glyptodont not protected by armour would have been the front of its face and its legs. Glyptodonts may have been able to bend down to shield their legs and it seems likely that, when faced with a predator, they would have turned away and presented their heavy tail as a deterrent.

In several South American species, the end of the glyptodont's tail resembled a heavy, mace-like club. This was undoubtedly used for defence. From studying marks on the shells of these species, it is clear that glyptodonts also fought each other, possibly over territory or females. One can only wonder at how they mated.

Nine-banded armadillos are famously clumsy but they can run and dig with incredible speed.

BIZARRE AND BEAUTIFUL ARMADILLOS

Although not directly related to the glyptodont, there is one mammal alive in Florida today that possesses body armour – the nine-banded armadillo. Despite being active mainly at night, this unmistakable and bizarre creature is familiar to many people. Famous for its poor hearing and eyesight, it has to rely heavily on its sense of smell and spends a large part of the time nose to the ground searching for the insects that make up the bulk of its diet. Armadillos often appear incredibly clumsy and almost unaware of what is around them, including other animals. However, when startled they leap straight up in the air. Although highly amusing to us, this is often enough to scare away would-be predators.

On close inspection, the body armour of the armadillo is not very thick; it is more like a tough, leathery coating. This allows the animal a great deal of flexibility. While it may not stop the teeth of a large carnivore such as a black bear, it does make armadillos harder to get hold of. It also allows them to escape unscathed through even the thickest and thorniest of scrub.

Although they occur naturally in South and Central America and as far north as Texas, the nine-banded armadillo is not native to Florida. A few were introduced there in the 1920s and 1930s and in just a few decades they were common throughout the state. This is a testament to their adaptability and shows the decrease in their natural predators, such as the red wolf and coyote.

During the Ice Age, the beautiful armadillo lived in the region. 'Giant' might have been a more suitable name because this mammal is thought to have weighed several hundred kilograms and measured almost 3 m (9.8 feet) in length. The nine-banded armadillo is small by comparison, weighing a maximum of 8 kg (17.6 pounds).

DEATH OF A GLYPTODONT

There was one crucial difference between the armadillo and the glyptodont: the scales of a glyptodont formed a tough, rigid and inflexible armour. These defences were impressive, but one fossil find shows that they were not totally impenetrable.

The remains of several glyptodonts have been found in Safford, Arizona. One particular skull shows the unmistakable signs of a violent death. On the top of the head are two elliptical holes, each measuring 2 cm by 1.5 cm (0.8 inches by 0.6 inches). Clearly a large carnivore with huge canines had attacked this glyptodont, but how did it break through its armour?

A closer look at the fossil skull provides the answer. It is slightly smaller than many other glyptodont fossil skulls, suggesting that this animal may have been a juvenile. Armadillos also possess head shields that are much softer when the animals are young. Only when they are fully grown does the skin toughen up. It seems likely that this was also the case with glyptodonts. The bony shield would have been only partly fused – no match against the teeth of a large hungry predator. So, the next question is, which predator was the most likely culprit?

Left: Rarely glimpsed, the enigmatic Florida panther was declared the state animal in 1982. **Right:** The flooded swamps of the southeast are home to a huge variety of animals. Here, a raccoon feeds cautiously near an American alligator.

Today only one large predator lives in Florida – the Florida panther. This enigmatic cat grows up to 2 m (6.5 feet) long, with males weighing up to 70 kg (155 pounds). The panther is a subspecies of the puma or mountain lion, which lives in the western part of the continent. Panthers are pale brown or rust-coloured and often have faint spotting in their coat. They are generally slimmer than their western cousins.

Panthers feed mainly on white-tailed deer, feral wild hogs and raccoons. A single animal may have a home range of up to 1000 sq km (386 sq miles). Until recently they could be found throughout the southeast, but now the only self-sustaining population is in southern Florida, notably the Big Cypress Swamp. Today, Florida panthers are one of the rarest species in North America, with maybe only 50 animals alive. They are seldom seen, even by the scientists who study them. The panther covers up the remains of its prey to avoid detection.

Thirteen thousand years ago the panther was much more widespread, and there were many other, far larger, predators around at that time. These included the Florida lion, possibly the largest cat ever to have lived, and the most notorious of all the big cats – the sabre-tooths and scimitars, which were known to attack prey much larger than themselves.

Another fearsome predator was the jaguar. These exquisitely coloured cats have a yellow-orange coat, intricately dotted with complex black spots. As recently as the nineteenth century, jaguars were known to live and breed in Arizona but there have only been a handful of sightings here during the last 150 years. Jaguars are now found in Central and South America. From nose to tail, males can reach over 2.5 m (8 feet) long and weigh over 150 kg (330 pounds), making them the largest cats in the Americas. Indeed for its size, the jaguar is probably the world's most powerful cat. However, fossils show that during the last Ice Age Florida jaguars were an even more impressive size – on average 20 per cent bigger than the largest jaguars alive today.

Although they will eat a great variety of prey, studies show that jaguars mainly feed on the biggest common animal in their territory. In South America, this is usually tapir, capybara or peccary. Jaguars have a distinctive way of killing their prey. Most large cats, having first overpowered their victim, finish them off with either a neck or muzzle bite. This is designed to suffocate their prey by blocking its mouth and nose or by crushing its windpipe. Jaguars, however, go for the head. Their coup de grâce is a bite to the skull, into the brain. They have an incredibly powerful bite and jaws hinged well back on the skull, which allow

Left: Weight for weight, the jaguar is probably the world's most powerful cat. **Centre:** Capybara, now found in South America, are the world's largest rodents. They can weigh up to 54 kg (120 pounds). **Right:** Tapir living in Ice-Age North America would have been one of the jaguar's main prey.

them to open their mouths very wide. So, its seems that the jaguar was the most likely perpetrator of the holes in the young glyptodont's skull.

THE SOUTH AMERICAN CONNECTION

Why did Florida have such a rich fauna, and what accounts for the variety of animal species present? The region was relatively warm during the cold Ice Age and it provided a refuge for many species from cold climates. Two different groups of animals were pushed together at this time: southern species, including warmth-loving animals from South America, and more temperate northern animals, forced southwards by the advancing ice sheets. In the southeast, north met south and this created an assembly of animals and plants quite unlike anything we can see today.

The Virginia opossum is one of the southeast's toughest creatures. Not the most choosy of animals, it will feed on almost anything and will even kill and eat poisonous snakes as it seems to be immune to snake venom. This highly adaptable creature, which is no bigger than a domestic cat, is equally at home

The Virginia opossum is the only marsupial mammal in North America. It's prehensile tail makes it very good at climbing.

feeding in forest ravines, scavenging on road kills or rummaging through garbage bins deep in the city. Opossums are found throughout most of the eastern United States. However, their range is limited by snow and frost. They do not hibernate and are not designed for cold weather. In the northern part of their range, their hairless ears and tails suffer from frostbite.

What sets these animals apart from all other mammals in North America is the fact that they are marsupials. Their young are born prematurely, and for two months females carry up to 13 babies in their pouch. These marsupials are a distinctly South American group of animals, and opossums are in fact the only marsupials found north of Mexico. They are living evidence of a migration of South American species. Before the Ice Age, many groups of animals expanded their range across the isthmus of Mexico and Panama and settled mainly in the southern part of North America.

Opossums are one of the few surviving species in North America with origins in Mexico and South America. However, the fossil record shows that many more southern species were once present, including the glyptodonts and capybaras already discussed. The most amazing of all these migrants – the ground sloths – were every bit as strange in appearance as glyptodonts.

GIANT GROUND SLOTH

During the last Ice Age there were four species of ground sloth in North America (see Chapter 4, p. 106). Most species were found throughout the continent, but one species was found almost exclusively in Florida and along the Gulf Coast. This was the spectacular giant ground sloth.

In 1975, the remains of one of these Ice-Age beasts were excavated from a pond near Daytona Beach. Huge elephant-sized bones, quite different from those of mammoths or mastodons, were brought to the surface. The vertebral column was nearly 4 m (13 feet) long, the skull over 60 cm (2 feet) long and the hind foot measured an astonishing 1 m (3.3 feet) from heel to toe – perhaps the biggest foot of any mammal.

However, most striking of all were the animal's claws. On its front feet were two claws, unlike the four claws of the other ground sloth species. In keeping with the rest of the animal, they too were huge, measuring up to 40 cm (16 inches) long. In life, they would have been covered in a bony sheath similar to human nails, making them even longer. Indeed they would have been so long they would have impeded the sloth's ability to walk. Instead, their feet would have been turned on their side, resulting in a walk unlike that of any animal alive today. Their teeth revealed a diet that was almost exclusively vegetarian. So these fearsome claws would have been used for grabbing branches and presumably for defence, although few predators would have even come close to such a huge animal.

Scientists have calculated that giant ground sloths weighed around 4000 kg (3.9 tons), joining those Ice-Age heavyweights the mammoths and mastodons. Remarkably, studies of their hip bones and strong tails show that they could rear up on their hind legs, like a grizzly bear. This would have enabled the sloths to reach high up into the top branches of trees – perhaps over 4 m (13 feet) above the ground, the same reach as a giraffe.

Also like a giraffe, the shape of the skull indicates that the sloths probably had a long tongue that would have been used to wrap around leaves. The nasal cavity was well developed so they may have had a good sense of smell. The teeth and remains of dung from South American fossils show that they were probably not fussy eaters, chewing almost all parts of the treetops, leaves, fruit and twigs.

GOING SOUTH FOR THE ICE AGE

The massive sheets of ice that covered the north of the continent forced drastic changes on its landscapes and habitats. As the ice sheets advanced, creatures not adapted to the cold were forced southwards. This had a domino effect across the whole of the continent, and even in the far south temperatures were lowered slightly, allowing hitherto more northerly species to survive there and pushing out species requiring warmth. As the Ice Age ended and conditions began to warm, many animals gradually left their shelters and migrated back north. Fossils of muskrat, for instance, have been found in Florida, whereas today these creatures live much further north.

Certain places, however, have retained their Ice-Age characteristics and are home to species that stayed put. Today, the copperhead snake is found in just one location in Florida – the bluffs and ravines along the Appalachicola River. The valleys here are filled with woodland quite unlike anything else in the state. In fact they are more similar to the hillsides of the Appalachian and Smokey Mountains, hundreds of kilometres further north. Because of their relatively cool climate, these valleys still hold remnant populations of species that migrated from the north – Ice-Age refugees hanging on.

Left: The copperhead snake is one of five venomous snakes found in Florida. **Right:** The gopher tortoise is a keystone species; its burrowing activity is essential to the ecosystem and provides refuge for almost 100 other animal species.

TORTOISES AS THERMOMETERS

Little Salt Springs in southwest Florida is a cavernous sinkhole over 60 m (196 feet) deep. Many Ice-Age fossils have been found here, but one in particular tells us with some degree of certainty what the climate was once like. The collapsed remains of a giant land tortoise were discovered on a ledge 26 m (85 feet) underwater. It measured nearly 3.5 m (11.5 feet) in diameter and was 12,030 years old. We can learn much about how this giant land tortoise lived by looking at living giant tortoises on the Seychelles and the Galapagos Islands.

Tortoises are cold-blooded and unable to stand the cold; they rely on the environment to maintain their body temperature. Florida's largest present-day tortoises, gopher tortoises, have to dig deep burrows to survive the occasional cold winter day. The colder the weather, the deeper these tortoises dig. In colder Utah, the northernmost gopher tortoises dig burrows about twice as deep as those in Florida. However, giant tortoises are unable to dig burrows – a clear indication that although temperatures in the north were icy cold 13,000 years ago, the winters in the southeast were actually milder than today. Florida was indeed a refuge from the cold.

MIGRATING MASTODONS

Just below the mastodon remains found in the depths of the Aucilla River lies a distinct layer of compressed plant material. Closer examination reveals that this layer is made up entirely of woody vegetation chopped into 0.5–1 cm (0.2–0.4 inch) pieces. These lengths match the cutting crests on mastodon teeth perfectly, proving that this layer is in fact mastodon dung. This tells us not only what the mastodons were feeding on, but also what plants were present 13,000 years ago. They included grape vines, white oak, cypress, hickory, beech, hazel and gourds.

Moreover, a closer inspection of the mastodons' teeth reveals some remarkable facts. The teeth built up layers of enamel as they grew, with each layer containing chemicals taken from the animal's diet. There were two chemical signatures present in the mastodons' teeth, which proved to have come from very different geographical areas. One of these chemicals was easily traced to the plants growing around the Aucilla River. To find the other one, scientists had to look further afield. The nearest place the foreign isotope could be found was 320 km (200 miles) north in Georgia. The only explanation for this is that these animals were spending alternating periods of time at each location. In other words they were migrating – a round journey of 640 km (400 miles) each year.

Similar migrations occur today. In America many animals, notably birds, migrate southwards to escape the cold and spend the winter in Florida. So perhaps thousands of years ago mastodons were moving in a similar direction. However, looking at the animals' preserved dung suggests a different story. Many of the plants they were feeding upon around the Aucilla River were fruiting at the time they were eaten. As these are summer fruits, the mastodons must have been in Florida during the summer months. In other words, they migrated northwards for the winter – the reverse of many present-day migrations.

This may seem strange until we look at one of Florida's seasonal quirks. During the winter months Florida gets cooler but perhaps more importantly it also gets drier. Between October and April very little rain falls, especially in the southern parts of the state. With the exception of a few key areas, the Everglades almost dry out. Such winter droughts may have been longer and more severe during the last Ice Age. So could it be that a seasonal water shortage was forcing such migrations?

With winter approaching, this group of mastodons begins its long migration north.

THE GIANT QUESTION

One of the striking things about the amazing array of Ice-Age animals is that most of them were giants. Mammoths, mastodons and giant ground sloths were some of the largest land animals ever to have lived. The Ice-Age species of capybaras, llamas, camels, lions and bears were far bigger than their modern counterparts. So why were they so big?

Adaptation to cold environments is one reason. For mammals such as woolly mammoths, a larger size meant a relatively smaller surface area through which heat was lost. For carnivores, there were greater numbers of large prey animals around to eat and to support a large size, and being bigger may have been an advantage when competing against other animals at scavenging sites. For herbivores, a large size would have been a good defence against predators.

Perhaps a better way of understanding the phenomenon is to look at the question from another angle. These animals are only giants in our eyes. The question is, why did only the smaller animals survive?

Growing to a great size is an extreme adaptation, and whenever change and extinction occur, extreme adaptations suffer. Large herbivores that require huge amounts of food are poorly equipped to adapt to changes in climate and vegetation. Similarly, big carnivores that need more meat become susceptible to changes in prey density.

One of the most extraordinary over-sized animals was the giant beaver. Although similar in appearance to today's Canadian beaver, these giants were the size of a black bear. Their teeth alone measured 15 cm (6 inches) and looked like hippopotamus teeth. It is not known if, like the modern beaver, they chopped down trees. In giant beaver fossils the front incisors do not have quite the same cutting surface as the modern beaver's. But giant beavers did lead the same aquatic lifestyle and, at 2.5 m (8 feet) long, they would have made quite a splash.

THE VANISHING WATERS

Just off the Gulf coast of Florida, near Panama City, the ocean floor is covered with a fossil forest – evidence of woodland growing 20 m (65 feet) beneath the ocean surface. The unmistakable conclusion is that the existing seabed must once have been dry land. Most compelling of all, explorers are now beginning to find spear points, fossil mammoths and even old river channels deep under the waves.

The explanation for this is simple, but still incredible. At the peak of the Ice Age, so much water was locked up in the glaciers that sea levels were affected across the globe. In Florida the sea level was some 70 m (230 feet) lower than it is today and as a result the shoreline was over 160 km (100 miles) further west. Florida's landmass dramatically doubled in size.

The region's unique geography and porous limestone bedrock meant that when the sea level dropped, the ground water table dropped with it. Florida effectively dried out. Many springs and sinkholes previously filled with water, or overflowing into rivers and streams, were reduced to empty holes. The Everglades did not exist; this area was probably dry scrub. Many of the region's largest lakes, such as Lake Okeechobee, covering some 1800 sq km (695 sq miles) today, vanished completely. Florida was in near-drought conditions.

But the state did not dry out completely. The same rocks that drained water from the land also channelled it into natural springs. There would have still been places where water pushed back up to the surface. The springs were an oasis in an otherwise dry habitat and aquatic animals made their homes in them. The water would have attracted creatures from far and wide. But as well as being a source of life, these springs and waterholes would also have been a place of death, as they attracted hunters too. The remains of many animals that fell into the water became fossilized and preserved for thousands of years. This explains the riddle of why so many bones now lie buried at the bottom of Florida's springs.

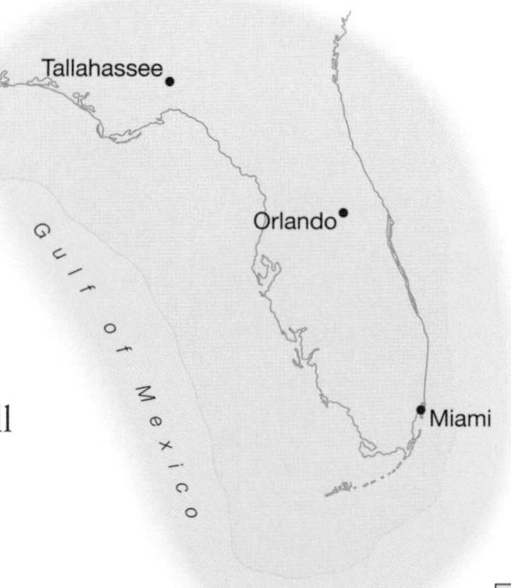

THE END OF A JOURNEY

The forces that parted the seas and lowered the water tables around Florida also created the land of Beringia from what is today the Bering Strait. This did far more than expose new land: it created a land bridge joining together the continents of Asia and America.

It was across this land bridge that the first people entered North America over 13,000 years ago as the massive ice sheets started to melt. Until this point, the continent had been untouched by man, yet in just a few hundred years people had managed to spread from Alaska in the northwest corner of the continent to Florida in the southeast: a distance of over 7000 km (4350 miles).

When people first arrived, they entered a virgin territory. The animals they encountered had no experience of humans, but that soon changed. The arrival of the first people coincided with the extinction of the vast majority of the big beasts described in this book. It was also a time of massive climate changes. What happened after humans arrived, and what was responsible for the mass extinction, is the subject of the final chapter.

KEY DEER - STRANDED BY THE ICE

Today the Florida Keys are a string of islands off Florida's southernmost tip. They extend in a chain 215 km (135 miles) long. At the peak of the Ice Age, these islands would have been hill tops in the middle of dry land.

As the ice sheets melted and the sea levels rose, the islands became isolated. One particular animal was marooned with no chance of escape – the Key deer. These animals are the Peter Pans of the deer world: they never grow up. Although they are the same species of deer as the white-tailed deer living on the mainland, they are considerably smaller. Mainland deer can weigh up to 135 kg (300 pounds) and grow to almost 2.5 m (8 feet) tall. Key deer, by contrast, never weigh more than 30 kg (66 pounds) or grow above 70 cm (2.3 feet) tall.

This miniaturization is a classic response to isolation on an island, and it can be seen in many animals living on islands around the world (see Chapter 3, p. 83). A smaller size may result from a limited food supply and a lack of predators.

Having completed their annual summer migration to Florida, two mastodons browse the rich foliage around a spring. In the foreground, a giant ground sloth rears up as high as a giraffe to reach the best leaves in a tree, while a glyptodont feeds in the shade below.

MAMMOTHS TO MANHATTAN

Left: Manhattan, New York: the ultimate symbol of modern North America. **Above right:** Mustangs are descendants of domesticated horses introduced by Spanish conquistadors. **Below centre:** Mammoths became extinct at the end of the last Ice Age just as the first people arrived in North America. **Below right:** A red-tailed hawk on its nest in downtown Manhattan.

Throughout this book we have looked at the landscapes and wildlife encountered by the first people to enter North America towards the end of the last Ice Age. But there is great controversy surrounding the question of who these first people actually were. In this chapter we explore this matter in greater depth and look at the way the arrival of our own species affected the continent's wildlife.

AND THE FIRST SHALL BE LAST

The Clovis First theory has for many decades been seen as the best explanation for how the first people entered the Americas. Our knowledge of Stone-Age cultures is based primarily on the tools they left behind. The Clovis culture is named after a town in New Mexico where one of the distinctive stone weapons that characterized this culture was first discovered. As more and more Clovis spear points were unearthed across the continent, the Clovis First theory proposed that humans arrived around 13,000 years ago across the land bridge that existed between Siberia and Alaska. These people then travelled through an ice-free corridor that opened up between the ice sheets in western Canada. In this way the Clovis people were the first to enter the interior of the continent. It is an elegant and almost biblical explanation. Nevertheless, to many it is too simplistic and almost certainly wrong.

One of the biggest challenges to the Clovis First theory has come from an unlikely location. In the late 1970s, archaeologists discovered an ancient campsite in Monte Verde in Chile, South America. This revealed a well-established community, thought to have been living in the area over 14,000 years ago. More than 20 years after its discovery, the accuracy of the Monte Verde find is still disputed, but it has raised the question of whether Clovis was in fact the first culture to enter the Americas. Based partly on the Monte Verde site and on other finds in North America, researchers are reviewing their estimate of when the first people arrived. Some suggest a more conservative figure of 15,000 years ago, while others are hedging their bets at 20,000–40,000 years ago.

Many now argue that, far from arriving through the interior of the continent on foot and spreading outwards to reach the coasts, the first Americans entered the continent via the coasts and pushed in towards the interior. Journeying by boat would certainly have been more rapid and may help to explain how people arrived at Monte Verde more than 14,000 years ago. As we saw in Chapter 2, there is some evidence to support the idea that early cultures utilized marine resources. This evidence is understandably patchy given that much of the coastline that existed in the last Ice Age is now buried under many metres of water. Nonetheless, clues to human cultures are turning up along the Pacific coast of Canada and America but, as yet, nothing has been discovered here that pre-dates the Clovis finds made in the interior.

TRACING THE FAMILY TREE

Any discussion about the first people raises the issue of where they came from. It is a sensitive issue because most modern-day native people of America and Canada believe themselves to be the direct descendants of the first people. Some argue that this is not the case and that modern-day native people are derived from later waves of migrants who came from Asia. Whether or not this is true, if Clovis were indeed the first people to enter the continent, then their origins lie somewhere in Siberia. Those favouring the Pacific coastal route theory have used

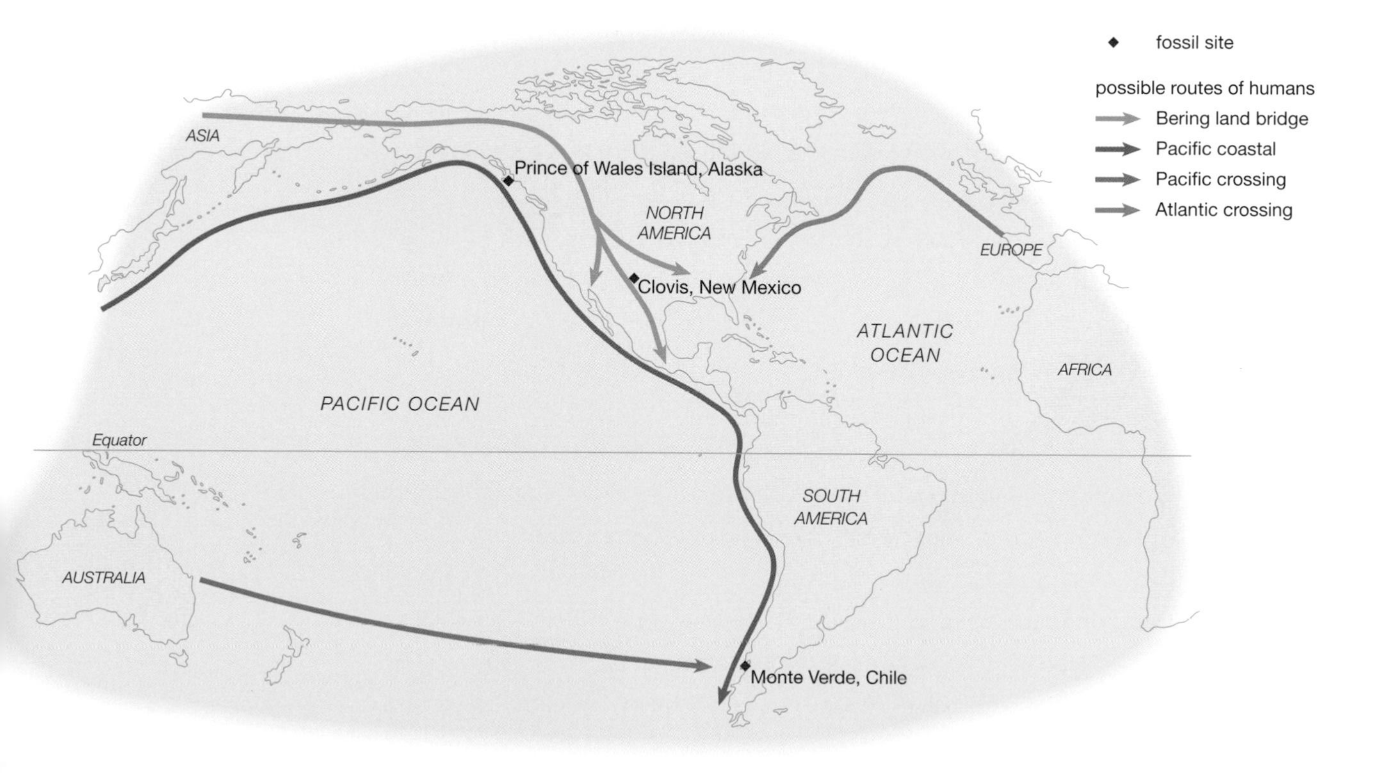

The first people were able to make fitted clothing and boots to cope with the extremes of climate at the end of the last Ice Age.

the dentition of human fossil finds to argue that northeastern Asia was the original home of the first North Americans. One radical theory throws the net much wider and suggests that their origin lay in southern France or Spain. This theory is based on the similarity between Clovis spear points and the Solutrean technology found in Europe between 16,000 and 44,000 years ago. It would have been an epic journey, requiring these first people to use boats to cross the Atlantic, perhaps following the edges of glaciers that covered much of the North Atlantic during this period.

So whether the first step into the continent was along a rocky coastline or across an open plain, it is hard to say exactly where the first North Americans came from. We can be certain that there were several waves of migration, possibly from many different cultures. Some cultures may have flourished and moved on; others may have died out or merged with different groups. Precisely when and how the first people arrived is not just an argument about who was first past the post – it helps us to understand more about the day-to-day lifestyles of these people.

ICE-AGE FASHIONS

Much of our understanding of the first people of North America, whether Clovis or pre-Clovis, comes from the stone technology they left behind. It is only recently that more attention has been given to what some call the 'soft technology'.

This refers to the perishable aspects of these cultures, such as tools and clothes made from wood and natural fibres. The classic image of hairy Ice-Age hunters wielding spears and sporting little more than a bundle of furry hides has, in recent years, been reinterpreted. The first people of North America were modern humans much like us. They communicated through language and were far removed from the knuckle-dragging cliché of Stone-Age man.

The Ice Age was not one long uninterrupted period of extreme cold. Seasons still existed and the first people would have been equipped for the extremes of winter and summer. Today the winters of the Alaskan interior see temperatures fall to as low as -50°C (-58°F). Similar conditions would have occurred in Beringia during the winter months. It is difficult to survive such cold conditions today, so 13,000 years ago the early North Americans must have worn something more substantial than a bundle of furs over their shoulder. There is evidence that they were making sophisticated seasonal clothing, using sinew from animals to sew leather hides into fitted outfits. During the summer months these first

ORAL HISTORY

Rather than digging around in the dirt for clues to the long-distant past, some researchers have taken a more lateral approach to the 'first people' debate. They have looked to language to provide important evidence.

The many different native peoples of modern-day North America speak more than 1000 different languages, which are collected into 150 language family groups. Using a mathematical model, in which every 6000 years a language splits into 1.5 languages, researchers concludes that it would have taken more than 30,000 years for the current number of languages to evolve.

The language debate has become a war of words in itself. Those believing that colonization of North America took place long before the Clovis people have welcomed the results. However, some linguists dispute the assumptions made in the model. They argue that while the number of individual languages may increase over time, the number of language family groups decreases the longer people are established. So although Africa has 2600 languages, it has only 20 family groups. The large number of native language family groups found in North America therefore suggests a shorter period of colonization – a completely opposite interpretation.

people had a new wardrobe. It is thought they wore lighter clothes woven from plant fibres, such as nettle, and possibly lighter footwear like sandals.

The cultural significance of this 'soft technology' has important implications for the debate on the origin and arrival of the first people. Those entering the continent along the coasts would have fished, foraged along shores and hunted sea mammals and birds. Nets and baskets would have been as vital as spears. Women, children and grandparents may have played an equal role in supplying food for the dinner table. Even if the first people were the Clovis hunters who entered the interior of the continent via the ice-free corridor, hunting big game would have only been one aspect of their culture.

Stone tools made up a tiny fraction of all the materials created by the Ice-Age people of North America. But while it is important to realize that these people were not slaying mammoths from dawn to dusk, we must not overlook the fact that they did have some highly sophisticated weapons at their disposal for hunting – not least the atlatl.

STONE-AGE KALASHNIKOVS

Imagine a device that enables you to throw a spear more than 200 m (660 feet) – twice as far as the best javelin throwers. The atlatl was such a tool. Derived from the Aztec word for 'spear-throwing tool', it actually originated in Europe, perhaps as far back as 25,000 years ago. An atlatl was a 60 cm (2 feet) long stick with a handle at one end for gripping and a hook at the other end to slot a spear or dart into.

Experts prefer not to call this weapon a spear thrower because it was probably used to launch darts. A dart was typically more flexible than a rigid spear and consisted of two parts: the main shaft, which was about 1.5 m (5 feet) long, and a foreshaft, which had a spear point lashed on with pitch and animal sinew at the end. The foreshaft was attached to the main shaft using a wooden or bone socket, and the total length of the dart was about 2.2 m (7.2 feet).

The dart was held in place by the hooked end of the atlatl and was supported between the forefinger and thumb of the thrower. It was launched in much the

The atlatl was a lethal Stone-Age weapon; it enabled human hunters to launch darts over a long distance with a high degree of accuracy.

same way as a baseball is thrown, with a flick of the wrist at the end of the throw. Although the dart could be projected over a distance of 200 m (660 feet), hunters would generally hit prey about 30–40 m (100–130 feet) away. The atlatl would have enabled the skilled hunter to hit a target with a greater degree of accuracy than simply launching a spear unaided. It also provided greater security by increasing the distance between hunter and prey. The atlatl was probably one of the most effective weapons used in hunting the big game of North America, including mammoths.

BRINGING HOME THE MAMMOTH

For most of us, the only danger in bringing home food for the family comes from kamikaze supermarket trolleys. The task facing the first people of North America at the end of the last Ice Age was a little more daunting. Armed with nothing more than spears and courage, they were faced with the prospect of killing a 5000 kg (4.9 ton) mammoth for dinner. But, the evidence for how they hunted is limited, with only 30 or so archaeological sites showing human/ mammoth interactions.

Most of the evidence for mammoth hunts comes from single animals rather than herds. In Naco, Arizona, the bones of an individual mammoth were found with eight spear points, most of them concentrated around the ribcage.

Other fossil finds suggest the slaughter of complete family groups at one time. At Dent in Colorado the presence of a group of mammoths at a cliff base may have been the result of hunters stampeding the animals over the cliff. Landforms could also have been used to trap mammoths, as at Colby in Wyoming. Here, a group of mammoths was killed in a steep-sided valley or arroyo. It is thought they were unable to escape because of drifted snow. Whether or not they were killed together or over a period of time is still uncertain.

Mammoths were not hunted just for their meat. Mammoth kills were also a valuable source of fat. Rich in vitamins and energy, the fat of a single mammoth could feed a group of more than 20 people. Groups of mammoths were rarely hunted and may have been more selectively butchered than solitary kills.

Our best guess as to how these people hunted comes from an understanding of their technology and of mammoth behaviour, based on modern-day elephants. Clovis people were the finest hunters of their time, their weaponry the result of thousands of years of refinement. Although the Clovis spear point was a lethal weapon, which almost became an extension of the hunter, mammoths would not have been easy prey.

A hunting party probably consisted of between 2 and 10 men. There may have been a great deal of prestige attached to hunting mammoths, particularly if hunting a mammoth group. Experienced hunters

Hunters, acting in pairs or small groups, probably killed mammoths using atlatls.

respect and understand their prey and Clovis hunters would have been alert to the behaviour of mammoths. Like modern-day elephants, mammoths lived in herds based around a matriarch, while males lived in loose association with other bulls. A 5000 kg (4.9 ton) mammoth might appear to be an obvious target but, at the same time, it was a dangerous one. Hunters would have minimized the risk of being detected by keeping downwind of the animal as they surrounded it. One member of the hunt probably acted as a decoy, attracting the attention of the mammoth, while the others moved in closer for the kill.

Mammoth hides were about 1.25 cm (0.5 inches) thick, with fur on top. A good deal of force would have been required to thrust a spear through this layer. Throwing a spear from a distance is effective but there is the problem of accuracy The atlatl overcame this by allowing the hunter to throw a dart over a longer range, at a higher velocity and with a greater degree of accuracy than a hand-held spear. By aiming at the ribcage there was a good chance of hitting the lung cavity and delivering a fatal wound. Even so, a wounded mammoth could take hours or even days to die, and the hunters would have had to track it until its eventual death.

THE END OF AN ERA

In the last two million years, the Earth has undergone more than 20 Ice Ages. It has been a time of extraordinary turmoil for the animals and plants living in the shadow of the glaciers. For the most part they have adapted to these changes by moving to new habitats as the ice sheets advanced and retreated. In North America only a handful of species went extinct during this incredibly long period.

It was not until the very end of the last Ice Age that large numbers of species suddenly disappeared. Most of these were big mammals weighing over 450 kg (990 pounds) – the so-called megafauna. In as little as 2000 years (the blink of an eye in geological terms), between 13,000 and 11,000 years ago, around 70 species went extinct in North America. They included mammoths, mastodons, ground sloths, sabre-toothed cats, camels and horses. Extinction was not limited to the larger mammals – some birds, reptiles and small mammals became extinct too – but the close of the last Ice Age will be remembered for the dramatic disappearance of the continent's megafauna.

The cause of the extinctions has fascinated scientists since the discovery of the first mammoth fossil 200 years ago. Today, the arguments for the post-Ice-Age extinctions of North America fall into three broad camps – overkill by humans, rapid climate change and death by disease.

BLITZKRIEG

For the last 30 years, one particular argument has caught the imagination of both scientists and the public alike. This is the 'overkill hypothesis', often referred to as 'blitzkrieg'. It points a very damning finger at the impact of humans.

The extinctions coincided with the arrival of the first people over 13,000 years ago. Well equipped to hunt, these humans came across animals that had never encountered people before. Successive generations of hunters moved south-wards, covering more than 300 km (186 miles) with each generation. In little more than 500 years, they could have reached the southern tip of South

America. Like the advancing front of an army, these people hunted and decimated many of the large herbivores and upset the ecological balance. Without the horses, mammoths and mastodons they fed upon, many predators and scavengers (including the sabre-toothed cat and dire wolf) disappeared soon afterwards. The swift movement of hunters across the continent explains the speed of the extinctions. It is easy to see why the overkill hypothesis has been nicknamed 'blitzkrieg'. But could Stone-Age hunters really have wiped out many of North America's biggest animals?

As we have seen, evidence of people hunting large mammals does exist. The presence of spear points among the fossilized bones of mammoths and mastodons is often seen as the 'smoking gun' of human-led extinction. The overkill hypothesis is further supported by similar extinctions that coincided with the arrival of humans on other continents. Australia, for example, lost all of its very large mammals (including giant wombats and giant kangaroos) around 50,000 years ago – not long after it is believed the first people reached its shores. In recent years, mathematical models have been used to examine the growth rate of the human population and the likely impact that hunting would have on prey species. They suggest that it was feasible for people to have wiped out the megafauna in as little as 1000 years.

However, despite apparent mathematical proof, the overkill hypothesis has increasingly been seen as simplistic and lacking in hard evidence. Spear points have been found in association with mammoth and mastodon remains, but none have been discovered among the remains of other extinct herbivores such as horses and camels. In addition, there has been a reinterpretation of the radiocarbon dates for the extinction of many species. This has led some researchers to argue that many animals were on their way to extinction before people arrived. In fact, mammoths and mastodons may have been the last of the larger mammals to become extinct. Predators such as sabre-toothed cats might have died out before them, contrary to the 'overkill hypothesis'.

Finally, if a wave of people had spread southwards on foot across the interior of North America, the kills in the south should be more recent than those in the north. But fossil finds where spear points have been found among the bones show no such pattern – indeed many of the earliest kills come from the southern part of the continent.

A CHANGE OF CLIMATE

Extinction through climate change is a more complex, if equally problematic, alternative explanation to overkill. The argument goes that as the glaciers retreated and the climate began to get warmer and drier, many animals were unable to adapt and died out as a result. Of course, the influence of climate change is far more complicated and varied than this simple explanation allows for.

It is argued that the alteration in climate upset the long-established ecological balance that existed between plants and animals in their Ice-Age ecosystems. In some areas, this may have resulted in the total disappearance of well-established habitats, dramatically affecting the animals they supported. Breeding patterns could also have shifted with the changing climate and vegetation. Plant and animal species may have migrated across the continent to where new ecosystems were beginning to develop. Even where the changes were not as extreme as, for example, grassland turning to forest, it is thought that the nutritional values of habitats decreased when different plant species mixed. The nutritionally rich grasses of the Mammoth Steppe could have been replaced by plant species that contained toxic alkaloid chemicals. Large herbivores such as mammoths and mastodons would have suffered the most from the loss of nutritionally rich habitats. The complexity of these climate changes could account for the extinction of some animals and not others.

Opponents of the climate-change theory point out that there have been more than 20 other glaciations, which all the big mammals lived through unharmed. Nevertheless, we have little knowledge of what happened at the end of those previous Ice Ages and it is impossible to draw direct comparisons.

Some proponents of this theory argue that the influence of climate change would have been felt gradually, over many thousands of years. However, the evidence points to a rapid disappearance of the megafauna, with species becoming extinct in as short a period as 1000–2000 years. Nonetheless, the rapid rate of the North American extinctions may actually support climate change as the cause. It is possible that so many animals died out at the end of the last Ice Age because the change in climate was far quicker and more devastating than after previous Ice Ages. Unfortunately, there is such a paucity of information that nothing has been proved conclusively.

Climate change at the end of the last Ice Age may have been more sudden and dramatic than at the end of previous Ice Ages.

IVORY INDICATORS

Mammoth and mastodon tusks may hold clues to the cause of these animals' extinction. Tusks grow throughout the whole of the animal's lifetime in conical layers from deep inside the tusk socket. Like tree rings, these layers can tell us much about the development of the animal – its health and the impact of environmental stresses. The rings at the tip of the tusk represent the animal's youth, while those at the base represent the final stages of its life.

Analysis of elephant tusks provides a template for assessing mammoth and mastodon tusks. We know a lot more about elephant societies than their extinct relatives. When an elephant dies, the different stresses in its life are reflected in its tusk patterns. As they mature, male elephants are evicted from the family unit and this stressful time is recorded in the tusks. The tusks can also indicate the condition the animal was in at the time of death and what season it died in.

The application of this research to mammoths and mastodons is in its infancy. Early results suggest that, after the last Ice Age, some male mammoths were reaching sexual maturity at a younger age. A similar shift in elephant societies is usually a reflection of hunting pressures, rather than climate change. While not conclusive, this is an intriguing way of examining the impact of human hunters on mammoths.

Bacterial or viral disease, spread by humans, is one of many explanations for the extinction of animals at the end of the last Ice Age.

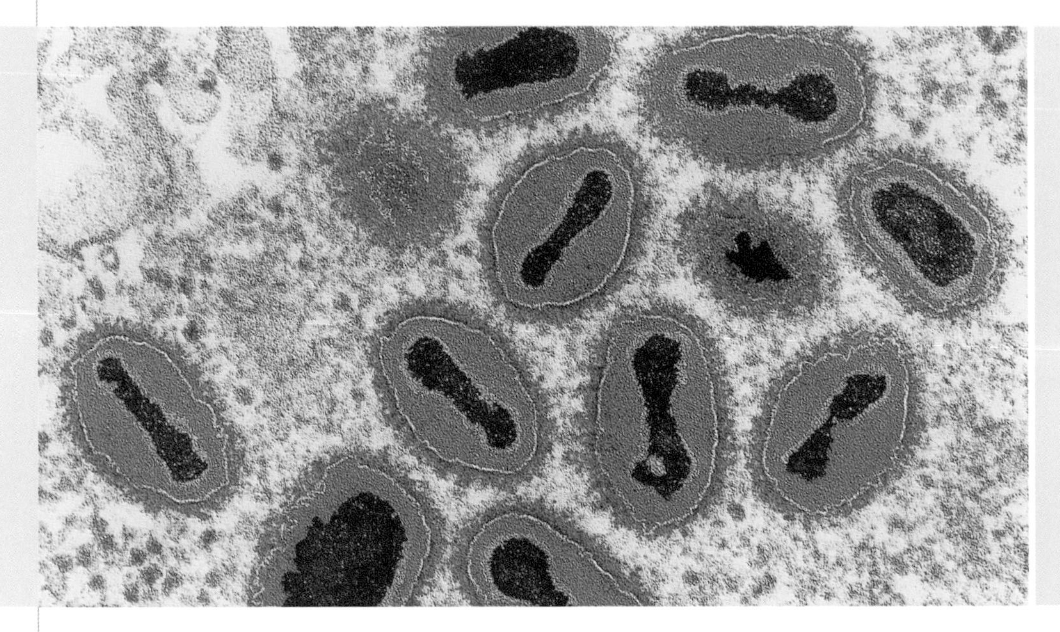

HYPERDISEASE

Extinction through disease is the most recent, and perhaps the most radical, explanation for the disappearance of so many species at the end of the last Ice Age. This argument proposes that a 'hyperdisease', similar in scale to AIDS, was carried into the continent either by the first people or by the early wild dogs living around, but not with, humans. The carriers of the disease would have some level of immunity, whereas the animals of North America would not. To cause extinction, the disease would have to have spread rapidly across the infected animal populations. Larger mammals would have suffered more, so the theory goes, because their slower reproduction rates would have prevented them from bouncing back quickly. If mortality rates had reached a threshold of 75 per cent the species would have been on the brink of extinction and would have been easily wiped out by other factors such as hunting or climate change.

The idea of a rapidly spreading and lethal disease is not without precedent – smallpox and measles, carried by European colonizers, wiped out many Native American communities from the early 1500s. However, the evidence for a disease-led extinction of the North American megafauna is, like the overkill hypothesis, based on the coincidence of human arrival. As yet no pathogen has been identified and many people remain sceptical that one can ever be found.

The extinction of the largest mammals is still a mystery. Most recently, a 'second order overkill hypothesis' suggests that people were indirectly responsible for the extinctions by hunting predators rather than herbivores in the first instance. Without predators, herbivores flourished, eating everything in sight. However, with no predators to exert control over herbivore numbers, they then suffered an inevitable population crash and became easy pickings for hunters. Ruminants like bison survived because they are more efficient at processing nutrients from grazing. This more involved version of the original overkill hypothesis has added to the ongoing debate.

For several decades the arguments over the Ice-Age extinctions have been dominated by the question of whether or not humans were responsible. By polarizing the issue along these lines, researchers have exhausted themselves trying to find the definitive answer. In reality, the answer is probably more complex than a simple yes or no. For North America, the end of the last Ice Age was a time of coincidence – the arrival of the first people, a massive change in climate and the disappearance of most of the continent's biggest animals. Perhaps it is time to accept that both people and climate had a hand in the extinctions.

THE MAMMOTH'S LAST STAND

The end of the Ice Age saw the disappearance of the mammoth from Europe, Asia and North America. It was thought that the last of the world's mammoths survived in North America until around 12,500–13,000 years ago. This view changed in 1993 when researchers found the fossilized remains of mammoth bones on Wrangel Island in the Chukchi Sea, north of Siberia. This proved that mammoths were living here as little as 4000 years ago. These animals are now believed to have been the last mammoths to walk our planet. It is extraordinary to think that mammoths were still around when the pyramids of Egypt were being built.

The mammoths found on Wrangel Island were at first thought to be dwarf forms. Dwarf mammoths are known to have existed on many islands, including the Channel Islands off the coast of California. Now researchers suspect that, though the Wrangel Island mammoths may have been slightly smaller than their mainland relatives, they were not dwarfs.

Advocates of the overkill hypothesis often quote the Wrangel finds as supporting their argument. They see the long-term survival of the mammoths here as a reflection of their isolation from human hunters. There is still speculation about when people arrived on the island, but it may have been at around the same time as the last of the world's mammoths disappeared, again supporting the overkill theory.

THE END OF THE LINE

The end of the Ice Age was marked by another significant extinction – the disappearance of the Clovis culture. Soon after the demise of the megafauna, Clovis stone tools no longer appear in the archaeological record. With no large mammals to hunt, these skilled hunters seem to have vanished from North America. It is uncertain what actually happened to the Clovis people, but new cultures began to appear around the same time as they disappeared. The ubiquitous Clovis culture may have splintered into localized cultures, each adapted to the unique conditions of the changing landscape. On the other hand, these new cultures may have arrived hot on the heels of the Clovis people and proved to be better adapted to the altered landscape. Whichever theory is correct, the Clovis people would have been absorbed into the burgeoning cultures of the New World, which were developing new relationships with the wildlife all around them.

THE SURVIVORS

Many animals flourished after the extinctions. Moose, previously limited to Beringia, spread southwards to occupy the browsing niche left by mastodons and the larger stag moose. The grizzly bear was one of the biggest surviving predators and its range expanded. Another survivor was the bison.

During the Ice Age, the bison on the great plains were much larger than modern bison and they formed smaller herds. The origin of modern bison is still uncertain, but a distant ancestor may have been the long-horned bison, which at twice the size of the modern bison had horns that spread 2.1 m (6.8 feet). It is clear that bison underwent a change towards the end of the last Ice Age, shrinking in size. They may have adapted in response to the available food supply. There was plenty of competition for food: at that time huge herds of horses were the dominant grazers on the open plains. However, the horse became extinct in the Americas at the end of the last Ice Age. Soon afterwards, bison numbers expanded and they began to form the large herds familiar to us in the cinematic image of the Wild West. The downside of the bison expansion was that these grazers became the focus of attention for some of the new hunting cultures on the plains.

Traditional gourds, hung from wooden racks, make perfect nesting sites for purple martins.

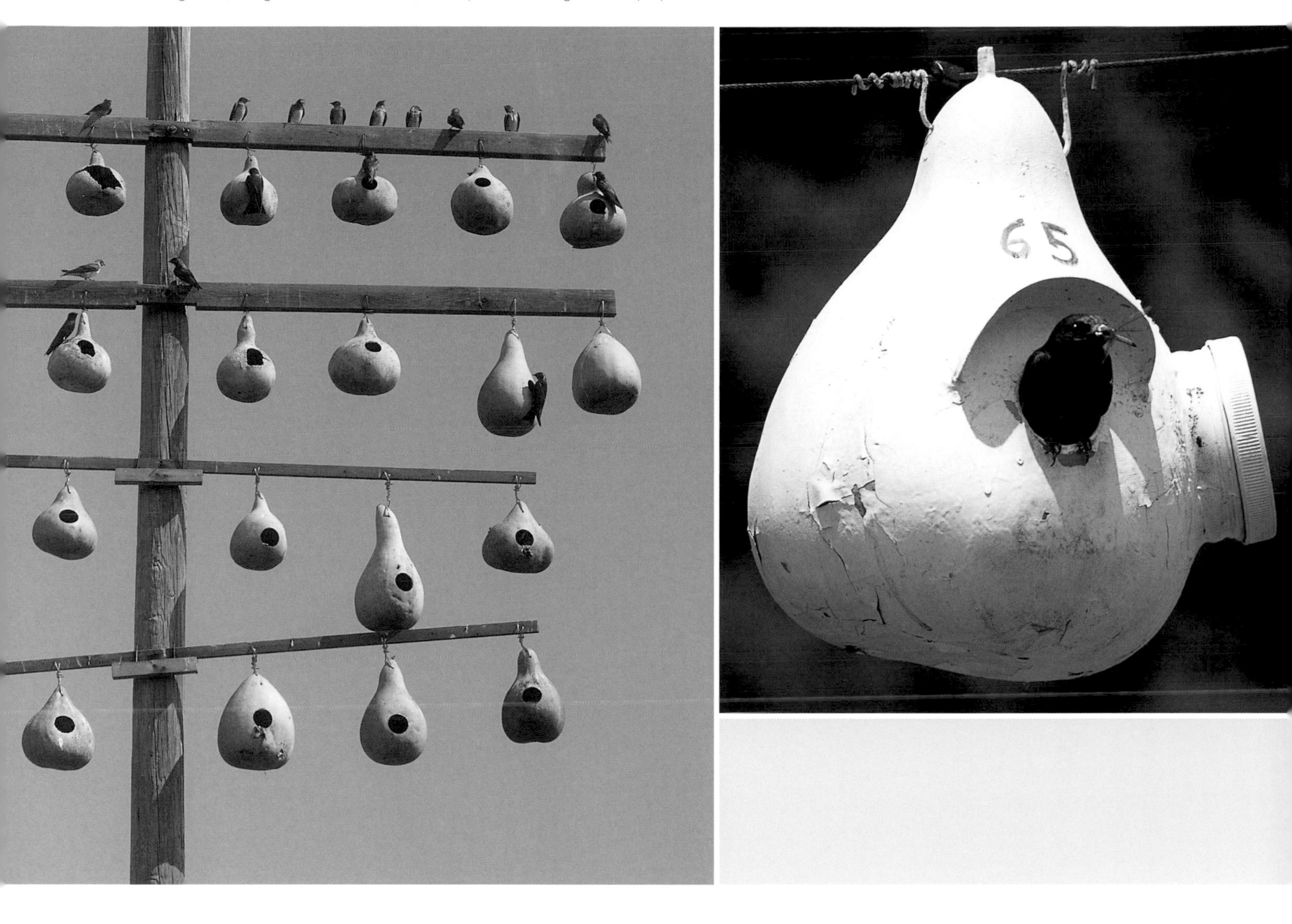

Some native animals benefited from the emergence of these cultures. The purple martin is found throughout North America. Before the first people entered the continent, the martins nested in the abandoned chambers left by woodpeckers or in holes in cliffs or trees. Today, purple martins east of the Rockies are dependent on nest boxes put out by people. It appears that this reliance originated thousands of years ago but how the relationship started is still uncertain.

Native Americans traditionally left gourds out for purple martins to nest in. Perhaps this started by accident, with a pair of martins making use of an empty drinking gourd beside a lake or pond. Whatever the origin of this tradition, native people began to hang out gourds from wooden racks, and this attracted

the birds to nest around their camps. The martins may have acted like a scarecrow, frightening off other birds from the corn patches. Alternatively, they might have fed on insects and pests like mosquitoes and midges. Perhaps the people simply enjoyed having the birds around, watching them flying and listening to them singing.

Not all visitors to the camps were as welcome as the martins. At least 3000 years ago, possibly even earlier, the first signs of agriculture appeared in North America, with some of the earliest crops being introduced by people from Mexico who had migrated northwards. Squashes, sunflower seeds and corn were all grown in small gardens. However, with agriculture came the first pests. Birds such as crows, grackles and redwing blackbirds all took advantage of the new crops. By the time the first Europeans arrived, Native Americans were clearing huge areas for crop planting. Pest birds were flourishing and, as a result, special raised platforms were built to allow for a constant watch over vulnerable crops.

The influence of native cultures on the landscape and wildlife has often been romanticized as a custodian relationship. In recent years, this view has been reassessed and it seems that the indigenous people of North America had a greater impact on their environment than previously thought. However the influence of people on the environment moved to another level when, 500 years ago, a new

Left: A nineteenth-century artistic depiction of Native Americans hunting bison on horseback. **Right:** Lithograph showing the conquest of Mexico by Hernando Cortés, 1519–21.

culture 'rediscovered' North America. A second wave of colonization began soon after the arrival of European explorers such as Christopher Columbus and John Cabot in the New World. This was to have an even more dramatic and rapid impact on the wildlife of North America than the changes that occurred towards the end of the last Ice Age. By the early 1500s, European settlers, especially the Spanish, were making their mark on the continent. They brought with them an animal whose ancestors once roamed wild across the landscape – the horse.

THE AMERICAN DINGO

The first people to enter North America encountered many animals they had never seen before. However, the woods of South Carolina hold clues to an animal that may have been familiar to the first people arriving in the continent over 13,000 years ago.

Many researchers believe that packs of dogs living wild in these woods are descendants of America's first dogs. In many ways, the Carolina dog resembles the Australian dingo. Its colour, size, shape and even its behaviour suggest it could be a close relative of primitive dogs. Some people now breed domestic Carolina dogs, although their breeding patterns are unlike most domestic dogs, and they will often dig out elaborate dens to give birth.

Genetic tests suggest that the Carolina animals are closely aligned with more primitive dogs. However, the results are far from conclusive. Early dogs are known to have lived around humans as long as 100,000 years ago. Far from being tame, they probably skirted around human camps, living off scraps though keeping a wary distance. Dogs may well have followed the first people across the Bering land bridge into North America, but the tame domestic companion did not appear in human societies until thousands of years later.

A HORSE FOR A BISON, A BISON FOR A HORSE

The story of the horse in North America is inextricably entwined with the rise and fall of the bison. The very first horses evolved in North America around 55 million years ago. Many species developed, yet at the end of the last Ice Age there were perhaps only four remaining. Then came extinction. The horse, which had been one of the dominant grazers on the great plains, vanished from North America, the land of its origin. Bison lived through the extinctions, perhaps better equipped to extract nutrients from dwindling food resources. Already shrinking in size during the Ice Age, they continued to do so afterwards. Nonetheless, the modern bison of the great plains was the largest grazer on the continent after the megafauna disappeared.

However, bison numbers did not immediately expand to take the place of horses. Between 9000 and 5000 years ago, rainfall decreased dramatically across the great plains and led to drought conditions. Still, the resilient bison survived, although their numbers declined. When, about 5000 years ago, the rains came back to the great plains, bison numbers exploded. Hunters regularly trapped and killed bison but, with few other predators, the animals thrived until their numbers reached mythical proportions. It is thought that roughly 65 million bison grazed the great plains at the time of the Europeans' arrival.

In the early 1500s the bison's luck ran out and its old nemesis – the horse – was back. Introduced mainly by the Spanish conquistadors, the horse was used during the Spanish invasion of Mexico under the command of Hernando Cortés. Cortés said of his victory over the Aztecs, 'Next to God, I owe it to the horses.' Not long after that, the horse spread across the continent with the colonists. By 1800 it had been adopted by the Native American cultures of the great plains.

The reintroduced horse was a domesticated descendant, far removed from the horses that once roamed wild on the continent. Its return was bad news for the bison. Native cultures and European settlers used the horse to hunt bison herds. Prior to the horse, native people had hunted on foot, and dogs were used as a means of transport. The horse enabled people to carry more and therefore allowed for greater trade and an increase in possessions. Whereas previously six or seven bison hides had been used for tepee construction, now larger tepees of up to 20 hides could be constructed.

It is difficult to determine the exact impact that hunting with horses had on the bison. What is certain is that the many different cultures of the plains were symbolized by the potent image of the bison-hunting Native American riding a horse. The growing European appetite for bison products accelerated the demise of the bison population. Some native cultures traded bison, but European settlers also hunted them. They used horses and a more effective weapon than a bow and arrow – the gun. Hunting expeditions took place on a grand scale and between 1872 and 1874 over 3 million bison were killed, only 150,000 of those by Native Americans. One hunter set a record of killing 100 bison a day for a month.

Left: Horses evolved on the American continent millions of years ago; the presence of mustangs on the plains today echoes this ancient past. **Right:** This pile of bison skulls provides a stark demonstration of the scale of their slaughter.

With the expansion of the railways across the continent, the pressure on bison numbers worsened. The average railroad boxcar could transport the bodies of 500 bison to processing plants. Moreover, the railways encouraged new settlers, and with them came an increase in farming. Cattle ranching expanded, robbing bison of their grazing lands.

The horse continued to add to the bison's demise. No sooner had domesticated horses arrived on the continent than they broke free, with many returning to the wild. By the late 1800s thousands of horses were living in herds on the plains – the first mustangs. Horses were well and truly back in the landscape. For the already troubled bison, this brought extra competition for grazing land and water resources. It was all too much. By the end of the nineteenth century, only 600 bison remained in the wild. The horse seemed to have pushed the bison to the brink of extinction. But in 1894 the bison became a protected species and today there are several hundred thousand wild bison, mainly living in American and Canadian national parks.

AMERICA'S FIRST BIG BIRD

While the Europeans brought domesticated animals with them, only one native North American animal was ever tamed – the turkey. Like bison, the wild turkey population was enormous at the time of the European colonization. There may have been as many as 10 million turkeys across the whole continent. Standing 1 m (3.3 feet) tall, with iridescent bronze feathers, wild turkeys are able to fly and can run at speeds of 64 km/h (40 mph). They have incredibly acute sight and hearing but a poor sense of smell. In spring, males, known as toms, display by fanning their tails, strutting and making the familiar gobbling call. They will often go head to head with other males to attract females.

The first domesticated turkey originated in Mexico. In the early 1500s, Spanish conquistadors sent tame turkeys back to Europe, where they caused a sensation, with roast turkey becoming a dish for special occasions. Colonists arriving in the Americas from Europe often brought tame turkeys with them. However, many preferred the taste of wild turkeys and began hunting them. Native Americans had hunted turkeys for thousands of years, but the birds were not highly prized and some tribes left the hunting to children. With the arrival of the European settlers, wild turkeys started to be heavily hunted. They became the centrepiece on the table at Thanksgiving and at other special occasions. Before long, turkey numbers went the same way as the bison. They were saved by a combination of events: sport hunters had to pay a tax, which ensured money went towards the conservation of turkeys, and laws were introduced imposing strict controls on hunting. The wild turkey population is once again buoyant, with more than 4 million across the whole continent.

COWBOYS AND COWBIRDS

Pioneer farmers and ranchers headed west across North America in the early to mid-1800s, seeking land to produce crops and rear cattle for the growing population of settlers. The completion of the first transcontinental railway in 1869 accelerated this expansion. The railroad opened up the land and this promoted development. The railroad was the future – the province of British Columbia agreed to join the rest of Canada in return for being connected by rail. Towns quickly grew as people migrated and land previously considered to be useless

A male turkey in full display. Once on the brink of extinction, the wild turkey is now widespread again across the continent.

wilderness was now turned over to farming. As the native wildlife of the continent was squeezed into smaller and smaller pockets of wilderness, the proliferation of the cowboy culture and their cattle benefited one particular bird – the brown-headed cowbird.

Before the European colonization of North America, cowbirds commonly fed around the bison herds of the west. These members of the blackbird family habitually pick off the insects disturbed by grazing bison. Cowbirds use songbirds' nests, with the mouse-brown female laying one speckled egg in each selected nest. In the past, cowbirds followed the bison as they moved across the open plains, and they laid their eggs in different nests as they travelled. A female cowbird can lay up to 40 eggs in a season and will divide her time between feeding and making repeated trips to potential nesting sites.

Two hundred years ago, cowbirds were restricted to the short-grass prairies of the northern great plains. However, European settlers soon changed that. As bison numbers decreased and farming expanded, cowbird numbers rocketed

because the birds took advantage of the feeding opportunities around groups of cattle and from waste grain in arable areas. Cowbirds spread eastwards and westwards. For the songbirds that were parasitized by cowbirds, the impact was devastating. In the past, cowbirds had followed the seasonal movements of bison, but cattle are more sedentary. Rather than laying eggs in selected songbird nests and moving on with the bison, cowbirds remained with the cattle and local songbird populations became adversely affected. It is a problem that persists today and it is compounded by the continuing loss of songbird habitats. Cowbirds are now nesting in songbirds' nests in urban areas too. In some years, nesting lazuli buntings in the city of Missoula suffer 100 per cent parasitism by cowbirds.

The cowbird may have benefited from the changing landscape of North America but many species did not. The wilderness was rapidly tamed, forcing animals to exist on the fringes of an ever-growing human society. As the population of North America grew, a new habitat appeared – the city. The emergence of cities was a mixed blessing for wildlife, and those species able to adapt to the changing landscapes took advantage of the few opportunities on offer.

Left: Cowbirds feed on insects disturbed by the movement of bison through the grass. **Right:** Cowboys and cattle replaced Native Americans and bison as the Wild West opened up to European colonization.

NEW HABITATS

The early towns of North America, particularly those in the west, had a real frontier feel to them. They were for the most part a small collection of buildings only a step away from the wilderness beyond. These towns have now evolved into huge cities, and the wilderness is usually a place that people travel to for a short vacation. Nevertheless, some modern cities still have a frontier feel.

Anchorage in Alaska is a modern, sprawling city, both a home to 250,000 people and to moose, grizzly bears and wolves. This is particularly true during the winter months when high snowfall forces animals further into the city centre. Moose can sometimes be seen holding up rush-hour traffic or nibbling on the plants of city gardens. Wolves are more wary of people, but they have been known to attack moose within the city. Anchorage is unusual because it is located in Alaska, the state with the largest remaining expanse of wilderness in North America. But wherever there are cities, wild animals will be found within their perimeters.

LIFE ON THE FRINGE

Urban sprawl is a disease of modern North America. Many wild animals live in ever-decreasing natural habitats that abut the suburbs. It is hardly surprising that the animals best able to survive these new environments are nocturnal, going about their activities when most of the human population is asleep.

Silicon Valley in California is one of the ultimate urban sprawls. Several small cities spread towards each other, creating a stranglehold around the remaining pockets of open land. Western burrowing owls are tenacious residents that make use of tiny strips of land yet to be built upon. Burrowing owls are ground-dwelling birds, but the western variety does not dig its own burrow. Instead it appropriates the burrows of ground squirrels. Somehow, this small owl survives among the gleaming office blocks and spaghetti junctions of Silicon Valley. Although they are seen during the day, burrowing owls become most active at dusk, hunting for small mammals and insects, often under the glare of street-lights. Adult owls regularly hover in mid-air before pouncing on their prey. Breeding pairs incubate 6–12 eggs and, in summer, fledgling chicks emerge from the burrow and sit at the entrance while their parents hunt for food.

As urbanization continues in the Silicon Valley area, burrowing owl numbers are falling. Further south at Bakersfield in California, another city-dwelling nocturnal animal has a rising population. San Joaquin kit foxes have flourished in the urban environment, while their rural relatives have become increasingly endangered. Bakersfield is a small city but one that is growing fast. The small grey, big-eared fox favours the fringes of the city, where development land is often left fallow for several years before being built upon. Kit foxes are believed to pair for life and they breed in December and January. Females give birth in underground dens with four to five pups emerging in April. It is not uncommon to see pups playing dangerously close to busy road junctions; cars are the biggest hazard for the urban foxes.

Bakersfield lies in an arid region of California and the many years of drought may be why foxes outside the city are suffering. Water conservation programmes and a good system of irrigation have created a more stable environment in the city. The foxes' main prey, the ground squirrel, is also abundant within the city limits. Fortunately for the survival of the species, the foxes' predators

Left: Moose and people are familiar neighbours in downtown Anchorage, Alaska. **Right:** San Joaquin kit foxes have flourished in the urban environment of Bakersfield.

such as bobcats and coyotes are uncommon. Nevertheless, the booming city fox population is threatened by two factors – the increased development of fallow land that is converted into suburban housing, and the ever-growing numbers of the ubiquitous, but non-native, red fox.

DEEP CITY DWELLERS

Most wild animals live on the fringes of cities or move between the city landscape and more rural areas. Few have become deep city dwellers, and those that have are often regarded as pests. Raccoons are one of the most successful city animals in North America. They are particularly adept at gaining access to waste, their nimble fingers easily opening the latches and handles on garbage bins. Toronto in particular has, in the past, suffered a problem with raccoons when these pest animals have taken advantage of rubbish thrown away by their human neighbours.

Left: Red-tailed hawks have made the luxury high-rise apartments bordering Central Park their regular nesting site. **Right:** Central Park is an urban space created for humans, which has greatly benefited wildlife.

Not all inner-city dwellers are seen as pests. Some animals adapt well in the city because they are encouraged to do so. As cities have become increasingly concentrated, the need for areas of recreation has led to the formation of city parks. The most famous is Central Park in New York. In 1853, 700 hectares (1730 acres) of undeveloped land were set aside to provide an area where the people of New York could enjoy healthy recreation. It officially opened in 1873 and since then the park has been an oasis for people and animals alike. More than 100 species of birds can be found across the park and thousands of migratory species pass through on their journeys north and south. As a powerful reminder that this habitat is within the heart of a city, the mockingbirds of Central Park can sometimes be heard mimicking the sirens of emergency vehicles.

It is not surprising that this sanctuary for wildlife should attract predators and, as is the way with Manhattan, the biggest hunter has become a local celebrity. Every year, a pair of red-tailed hawks makes an expensive Fifth Avenue apartment building its chosen location for nesting. Precariously positioned above a window on the top floor, the nest commands a good view over the park. Swooping over crowds of walkers, joggers and rollerbladers, the hawks regularly snatch squirrels or pigeons and carry them off to feed their chicks. Once the chicks leave the nest, the juvenile birds can be seen in the park, where they are photographed by tourists and locals alike.

The increasingly urban society of North America is far removed from the landscape that greeted the first people to enter the land thousands of years ago. Now, as then, humans threaten the existence of many of the animals that share their environment. We can no longer watch mammoths grazing on the great plains or witness sabre-toothed cats hunting camels in Arizona, but the human desire to connect with the natural world knows no bounds, and today North Americans can still catch a small glimpse of their prehistoric past in the most unlikely of places. In Florida, a popular theme park draws people away from the cities and allows them to take a safari through an African-style savannah. It is a manufactured world but, despite this, the experience is real. People stare in awe at the cheetahs, lions and elephants from the safety of their tour bus. The experience they get is more than just a taste of the exotic. With a little stretch of the imagination it is possible to picture the wildlife and landscape of North America as it would have looked to the first people arriving on the continent over 13,000 years ago.

ACKNOWLEDGEMENTS

This book owes its existence to the hard work and insights of too many scientists to mention here but the following were particularly helpful in our research: Kathy Anderson, Connie Barlow, John Byers, Ken Cole, Brian Cypher, E. James Dixon, Jim Dunbar, Scott Elias, Terry Fifield, David Gillette, Russ Graham, Erick Greene, Carl E. Gustafson, Dale Guthrie, Timothy H. Heaton, Andy Hemmings, Fritz Hertel, James R. Hill III, Mark Jorgensen, Mike Kunz, Adrian Lister, Dave Lovelace, Greg McDonald, Paul Matheus, Jim Mead, Bruce Means, Joe Muller, Blaine Schubert, Chris Shaw, Rick Sinnott, Alison T. Stenger, John Storer, David Webb.

However any errors in interpretation or fact are very much our own.

We are grateful to the FAUNMAP project and the Denver Museum of Nature and Science for their help in creating the distribution maps for the extinct species. Thanks also to the Alaskan Native Heritage Center.

The Ice-Age creatures and landscapes that have been recreated here are the product of the skill and talent of the BBC MediaArc team who worked on the television series. Our thanks go to the Design Director Alisa Robbins, Head of Animation Henry Lutman and the animators David Cox, Noel Mulvaney, Jeremy Horton, Francis Offei and Jessica Lee. Clym Sutcliffe then skilfully placed the creatures in their backgrounds on these pages.

Special thanks should go to Production Manager Liz Toogood and Production Coordinator Jenni Collie, both of whom worked hard to make our travel across the length and breadth of North America as easy as possible. John Brown was our imaginative principal cameraman, who brought his full creative skills to the task of reconstructing Ice-Age North America.

INDEX